Charles Darwin

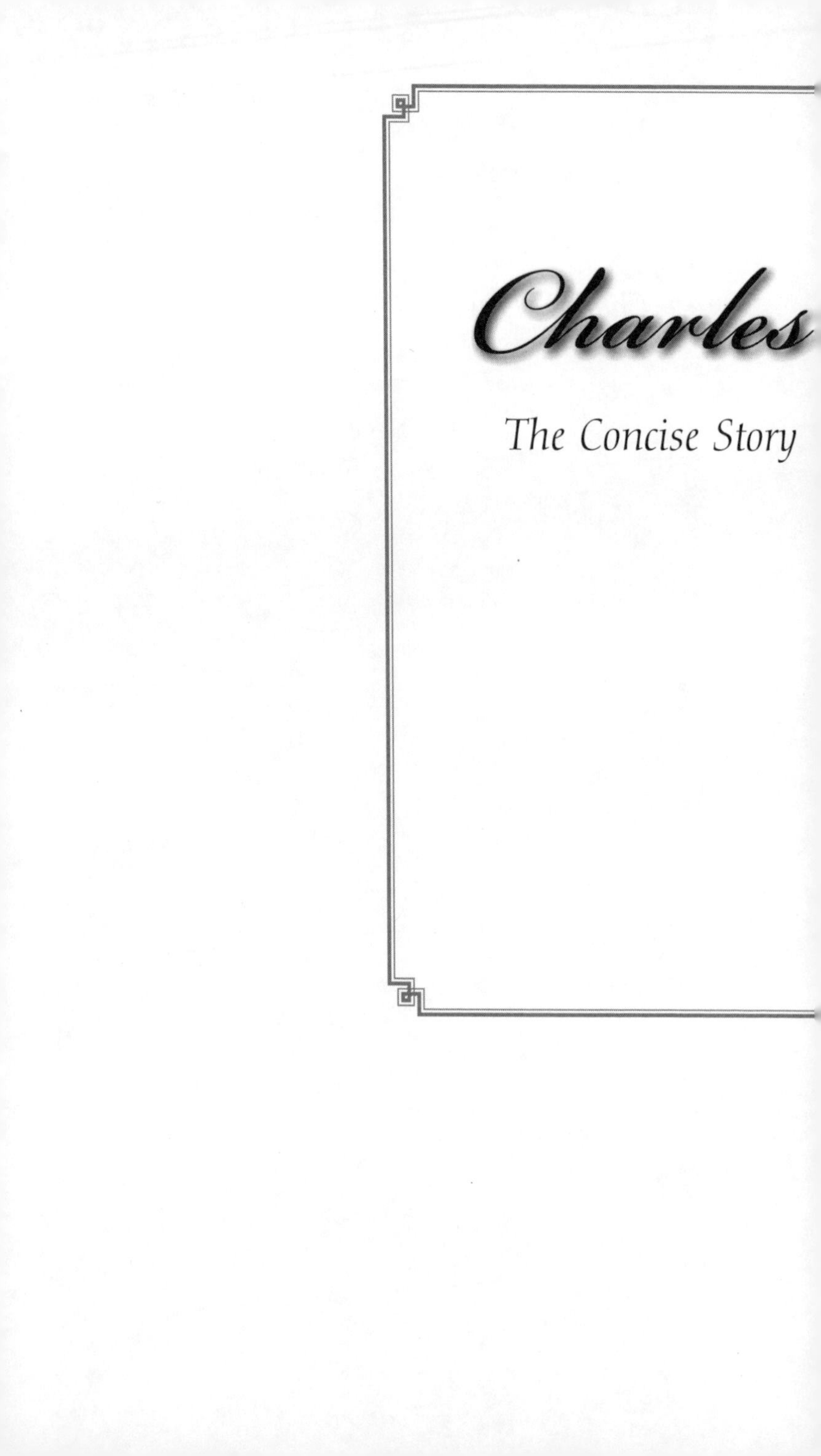

Charles

The Concise Story

DARWIN

of an Extraordinary Man

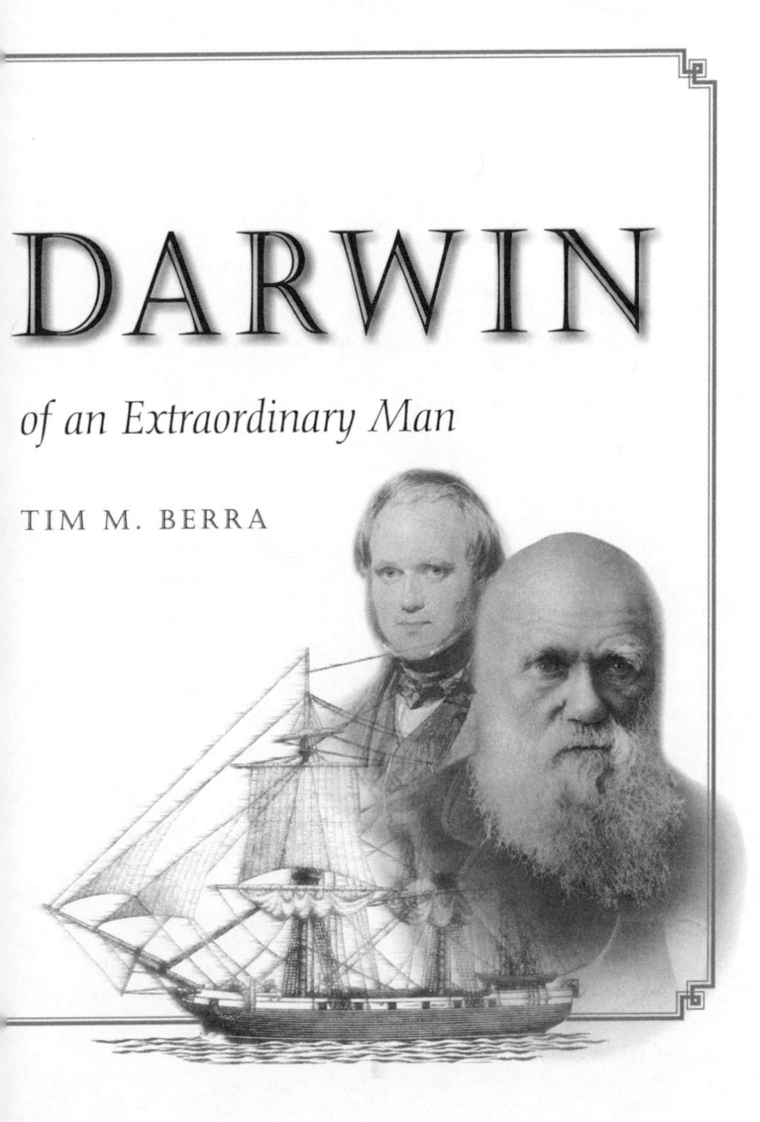

TIM M. BERRA

The Johns Hopkins University Press, Baltimore

Printed in the United States of America on acid-free paper
9 8 7 6 5 4 3 2

The Johns Hopkins University Press
2715 North Charles Street
Baltimore, Maryland 21218-4363
www.press.jhu.edu

LIBRARY OF CONGRESS CATALOGING-IN-PUBLICATION DATA

Berra, Tim M., 1943–
Charles Darwin : the concise story of an extraordinary man / Tim M. Berra.
p. cm.
Includes bibliographical references and index.
ISBN-13: 978-0-8018-9104-5 (hardcover : alk. paper)
ISBN-10: 0-8018-9104-3 (hardcover : alk. paper)
1. Darwin, Charles, 1809–1882. 2. Naturalists—England—Biography. I. Title.
QH31.D2B47 2008
576.8'2092—dc22
[B] 2008011320

A catalog record for this book is available from the British Library.

General note on illustration sources: Illustrations circa 1923 or earlier are in the public domain. Owners of illustrations are named, when they could be determined, at the ends of the captions.

Special discounts are available for bulk purchases of this book. For more information, please contact Special Sales at 410-516-6936 or specialsales@press.jhu.edu.

The Johns Hopkins University Press uses environmentally friendly book materials, including recycled text paper that is composed of at least 30 percent post-consumer waste, whenever possible. All of our book papers are acid-free, and our jackets and covers are printed on paper with recycled content.

To Judge John E. Jones III
and the late Judge William R. Overton,
whose decisions helped explain
why evolutionary biology is science
and creationism is not

Contents

PREFACE ix

Introduction 1

1. An Admirable Pedigree 2
2. A Privileged Youth 6
3. Exploration 13
4. Discovery 26
5. Maturity 34
6. A Proposal 38
7. Life at Down House 42
8. Correspondence 45
9. Daily Routine 47
10. Taxonomy and Selection 52
11. Alfred Russel Wallace and *The Origin* 59
12. What Darwin Said 68
13. Darwin's Bulldog 70
14. A Man of Enlarged Curiosity 72

15. Darwin's Death 81

16. Epilogue 85

APPENDIXES

A. Books 87

B. Chronology 93

C. Darwin Online 99

D. Dates 101

REFERENCES 103

INDEX 111

Preface

I HAVE BEEN LECTURING on the life of Charles Darwin for many years. These talks began as part of the introductory biology class I taught at the Ohio State University, and evolved into a public presentation as colleagues asked me to speak at their universities after my book, *Evolution and the Myth of Creationism*, was published in 1990. This lecture expanded as "Darwin Day" became more popular on university campuses, and I received more invitations.

After each talk, several people usually asked me if the text and illustrations of my lecture were published. They were, of course, but not conveniently in one place. The information is scattered throughout dozens of books about Darwin's life. What I offer here is not original Darwin scholarship, but a synthesis of existing essential information, along with a copious collection of illustrations conveniently gathered in one place.

My lecture was originally derived from Darwin's *Autobiography and Voyage of the Beagle*, R. B. Freeman's *Charles Darwin: A Companion*, Gavin de Beer's *Charles Darwin: A Scientific Biography*, and Alan Moorehead's *Darwin and the Beagle*. In 2000 I revised the talk, based on the obscure but wonderful little book *Down House: The Home of Charles Darwin*, edited by Louise Wilson and published by English Heritage. I have paraphrased this narrative, integrated it with my existing lecture, and augmented it with much additional information gleaned from many sources. For this reason I consider myself more of an editor or compiler than the author of this book.

The aim is to show not only that Charles Darwin was one of the most important men who ever lived but also that he was a good man—a decent human being who had a wonderful family life.

While polishing my lecture text for publication as this book, I visited the Darwin Exhibit at the Field Museum in Chicago in July 2007. At the museum's gift shop, in addition to a Darwin coffee mug, t-shirt, and model of the HMS *Beagle*, I purchased a little book in Oxford University Press's Very Interesting People series called *Charles Darwin*. It was authored by the three most important biographers of Darwin: Adrian Desmond and James Moore (*Darwin: The Life of a Tormented Evolutionist*) and Janet Browne (*Charles Darwin Voyaging* and *Charles Darwin: The Power of Place*). I mined this superb, brief account of Darwin's life for arcane tidbits that I had overlooked.

Anti-evolution fundamentalists have attempted to demonize Darwin and his magnificently useful insight, the theory of evolution. Hopefully, the reader will come to understand that "Darwin" is not a dirty word and that the theory of evolution is the best explanation that science can produce about the biodiversity of the natural world. It may very well be one of the greatest ideas anyone has ever had.

I AM GRATEFUL to the people who invited me to give my Darwin lecture and to the audiences who continued to press me for more details over the years. This especially applies to the audience in 2001 at the Northern Territory Museum in Darwin, Australia, whose standing ovation and thunderous applause were things professors rarely experience but secretly hope for in their classrooms. That stimulation and appreciation have allowed this small book to emerge. I appreciate the input of the anonymous reviewer who was clearly a Darwin scholar. I am grateful for the support of the Johns Hopkins University Press's Senior Editor Vincent Burke and for the thoughtful copyediting skills of Kathleen Capels. Thanks also to the people and institutions that allowed me to reproduce illustrations under their control. As always, I have had the support of my wife, Rita, who long ago accepted the fact of piles of books lying all over the house and has learned to vacuum around them.

Introduction

THE THEORY OF EVOLUTION is arguably the greatest idea the human mind ever had, and its proposer, Charles Darwin, is among the most influential scientists who ever lived. He changed the way humans view their place in nature. His explanation of the evolutionary process occurring through natural selection forms the basis of modern-day biological sciences, including the applied disciplines of agriculture, medicine, and, most recently, biotechnology.

There have been hundreds of books and articles about the life and times of this great man, and I obviously cannot cover everything there is to know in one short book. If you can read only one Darwin book, I highly recommend the 1991 biography by Adrian Desmond and James Moore; if you can manage two, add Darwin's autobiography. Here, I aim to present a more personal look at Darwin the man.

ONE

An Admirable Pedigree

CHARLES'S GRANDFATHER, Erasmus Darwin (1731–1802), was a physician and poet with a fascination for natural philosophy. He was popular as a doctor and did well financially because of his wealthy patients. Yet he refused payment from poor patients and often gave them money and food after he treated them.

One of his wealthiest patients and closest friend was Josiah Wedgwood, the famous potter who industrialized the manufacture of kitchen ware. They shared an interest in emerging technology, such as the steam engine, and both were members of a like-minded group of scientists, inventors, and intellectuals—including James Watt, who perfected the steam engine, and Joseph Priestley, credited with the discovery of oxygen—who met once a month, always during the full moon. This Lunar Society jokingly called themselves Lunatics. They opposed slavery, defended religious freedom, and supported American independence.

Erasmus Darwin proposed a natural explanation for the origin and development of life. In his 1794 book *Zoonomia*, he discussed the movement of climbing plants, cross-fertilization in plants, and the domestication of animals. In other works, he commented on the mechanism of inheritance and remarked on sexual selection. I mention these familiar Darwinian themes to suggest the sort of intellectual atmosphere

Dr. Robert Waring Darwin, Charles's father. This wealthy, prominent physician and Fellow of the Royal Society, at six foot two and 24 stone (336 lbs), was a dominant force in Charles's development. (Library, Wellcome Institute for the History of Medicine)

whirling around the household into which Charles's father and Charles himself were born.

Robert Waring Darwin (1766–1848), like his father Erasmus, was a widely respected physician, well connected with both the new industrialists and the local gentry. He was also huge, weighing about 300 pounds. Whenever Charles referred

Wedgwood anti-slavery cameo from 1791. Josiah Wedgwood distributed hundreds of these cameos to rally the populace to support the abolitionist cause. (Wedgwood Museum)

to his father, he added "the greatest man I ever knew," or "the kindest man," or some other sort of superlative. He had a great deal of respect for—and probably some fear of—his formidable father. Much of Robert's practice was psychiatric as well as medical. His patients held him in high esteem and valued his advice about their emotional problems.

Charles's mother was Susannah Wedgwood (1764–1817), Josiah's daughter, whose marriage to Robert was a result of the friendship between Erasmus Darwin and Josiah Wedgwood. Charles was born on 12 February 1809 (the same day and year as Abraham Lincoln), the fifth of six children (four daughters, two sons), into a privileged, affluent, and prominent family. His sisters were Marianne (1798–1858), Caroline Sarah (1800–1888), Susan Elizabeth (1803–1866), and his younger sister Emily Catherine (1810–1866). His older brother was Erasmus Alvery (1804–1881).

In a broader context, this was the era of King George III and Jane Austen. As in many similar upper-class households during this period, social life for the Darwins revolved around books, correspondence, conversations about literature and politics, and dinners with neighbors. The Darwins and the Wedgwoods had a great deal of respect for one another, and both families were involved in the anti-slavery movement.

TWO

A Privileged Youth

CHARLES HAD A PRIVILEGED and happy childhood at the Mount, the family home in Shrewsbury. The one exception was the death of his mother, Susannah, in 1817, when Charles was eight years old. The motherly devotion of his three elder sisters compensated for her loss, but his father became even more autocratic and overbearing after Susannah's death.

Charles escaped this patriarchal sternness with frequent visits to Maer Hall in Staffordshire, the nearby home of his mother's brother, Uncle Josiah Wedgwood II, and a multitude of cousins. After his mother's death, Charles was sent to boarding school in Shrewsbury. The school was only a mile away, so he walked the short distance home on weekends and, whenever he could, during evenings to collect beetle specimens or engage in chemistry experiments in the garden shed with his elder brother Erasmus (Ras), with whom he remained close throughout their lives. This hobby resulted in Charles's school nickname of "Gas."

Charles was unsuited to the classical education he received at Shrewsbury School (the only subjects taught were Latin and Greek, with a smattering of ancient geography and history) and was a mediocre student. This less-than-stellar academic performance displeased Robert Darwin. When Charles became bored with chemistry at age fifteen and took up hunting, his frustrated father erupted, saying, "You care

The Mount, the Darwin family home in Shrewsbury, where Charles grew up. It was built by Robert Darwin after his marriage in 1796. The photo dates from 1909.

for nothing but shooting, dogs, and rat-catching, and you will be a disgrace to yourself and all your family." Dr. Darwin then removed Charles from school and allowed him to serve as an assistant in his medical practice. Charles recorded medical histories from his father's patients and made guesses at diagnoses. This prepared him to join brother Ras at Edinburgh University in 1825, where he was the third generation of Darwins to have studied medicine.

Charles and Ras made the most of life as wealthy, young gentlemen in an exciting, big city. Charles thought the medical lectures were boring, and the horror of surgery without anesthesia was too much for him to take. In his second year at Edinburgh, Charles ignored his medical studies and concentrated instead on natural history. He visited local fishermen and sorted through their nets. He also gravitated toward zoologist and physician Robert Edmond Grant, a sponge expert who stimulated Charles's interest in marine invertebrates with field trips to the Firth of Forth, research projects, and presentations at scientific societies.

Charles attended a lecture by the visiting American painter and ornithologist John James Audubon and became fascinated with taxidermy. This led him to take private lessons on how to stuff birds from John Edmonston, a former slave who accompanied an expedition to South America and was probably the only black man in Edinburgh at this time. He loved Edmonston's stories of exploration, and the two became

Josiah Wedgwood II, Charles Darwin's "Uncle Jos" and father-in-law. It was Uncle Jos who persuaded Charles's father to allow Charles to go on the *Beagle* voyage. Josiah was the senior partner of Josiah Wedgwood & Sons, Ltd., and Emma's father. Painted by William Owen. (Wedgwood Museum)

friends. Charles was much more interested in natural history than in medicine, and he expressed his unhappiness in letters to his sisters.

In his second year at Edinburgh, Charles traveled to Paris, with his father's permission, to visit his cousins (including Emma Wedgwood) who were on holiday there. Charles and Emma, now both eighteen, had practically grown up together. In a letter to his sisters, Charles remarked on how beautiful Emma had become as a young lady.

Meanwhile, back at home, Charles's sisters prevailed upon Dr. Darwin to remove Charles from Edinburgh, which he did in April 1827. He decided that, if Charles was not going to become a physician, he should become a clergyman. This could only be done at an English university approved by the Church of England, so Charles was sent to Cambridge University to study for the church, even though the Darwins were not a religious household. Rather the opposite was true, as grandfather Erasmus, father Robert, and brother Ras were freethinkers.

Charles had some reservations about professing his belief in the dogma of the Church of England, but the thought of being a country parson appealed to him, as it would leave him time to study natural history. He needed to brush up on his Latin and Greek before he could go to Cambridge, so his father hired a private tutor. While studying, Charles discovered a new interest—girls. He became infatuated with Fanny Owen, a friend of his sisters. Apparently, all the young men of Shrewsbury fancied Fanny. Charles would visit her house, and they went hunting together. She was probably Charles's first girlfriend.

After eight months spent reviewing Latin and Greek, Charles was ready, and he left for Christ's College at Cambridge in 1828. This environment was much more to his liking. He engaged in beetle- and plant-collecting with like-minded students, including his cousin William Darwin Fox, who was his closest friend. He also fell in with the card-

Christ's College, Cambridge University, where Charles was a student from 1828 to 1831. *Inset,* William Darwin Fox, Charles's second cousin and beetle-collecting mate at Cambridge. (Cambridgeshire Collection, Central Library, Cambridge; Darwin Archive, Cambridge University Library)

Reverend Professor John Stevens Henslow, botanist and Charles Darwin's mentor at Cambridge University, by T. H. Maguire in 1849. (Fitzwilliam Museum, Cambridge)

playing, drinking set and did a fair amount of goofing off. He went home on vacations to see Fanny, but one Christmas he did not go back to Shrewsbury. Fanny was offended, and the flirtation cooled.

Fortunately, Charles came under the tutelage of Reverend Professor John Stevens Henslow, a botanist, who influenced his career more than any other person. Charles accompanied Henslow on field trips and was often invited to dine at his home. They took long walks together while Henslow instructed him in the natural history of the area; at Cambridge, Charles became known as "the man who walks with

Adam Sedgwick, professor of geology at Cambridge and Fellow of the Royal Society, who taught Darwin survey techniques on a field trip to North Wales in August 1831. (Drawing from Moorehead [1969], based on an 1855 photograph by Kilburn in the Hulton Archive.)

Henslow." He excelled in this environment and, unlike at Edinburgh, concentrated on his studies.

Henslow encouraged him to read widely, including Alexander von Humboldt's *Personal Narrative* about his 1799–1804 journey to South America. This inspired Charles to want to travel and filled him with a desire to contribute to natural history. Henslow arranged for Adam Sedgwick, professor of geology, to take Charles as his assistant on a summer field trip to North Wales and teach him the use of technical instruments, surveying, and geological descriptions. Charles was twenty-two when he graduated in 1831—tenth in his class of 178.

He arrived home at the Mount in August 1831 and found a letter waiting for him from Henslow. Inside Henslow's letter was another letter from Cambridge University astronomy professor George Peacock, offering Charles the opportunity for a voyage around the world.

THREE

Exploration

DARWIN REALIZED THAT THE OFFER to join HMS *Beagle* on a surveying trip circumnavigating the globe was the chance of a lifetime. The ship was a 10-gun brig, 242 tons and only 90 feet long. This class of ships carried the naval nickname of "coffin" because of their tendency to sink in rough seas.

The captain, Robert FitzRoy, wanted a gentleman-naturalist who would be a suitable traveling companion, and Henslow recommended Darwin as the best-qualified person for such an undertaking. Charles's father objected and produced a list of reasons why this "wild scheme" would not be a good idea. However, he did add that, if anyone of common sense whose opinion he valued thought it was a good idea, he would relent. As usual, Charles deferred to his father's wishes and rode off to Maer, twenty miles away, to hunt with Uncle Jos. When Charles explained the opportunity to his uncle, Jos quickly produced counter-arguments to Dr. Darwin's list of objections. He then rode back to the Mount and persuaded Dr. Darwin that this would be the perfect situation to help Charles settle down.

At their first meeting, FitzRoy almost rejected Darwin because of the shape of his nose. FitzRoy was a phrenologist who thought that Darwin's nose made him too weak for the exertions of a long sea voyage, but Darwin convinced him that "my nose had spoken falsely." Charles was about

HMS *Beagle* drawing by American artist Samuel L. Margolies (1897–1974). The *Beagle* was a sloop rigged as a brig. On her second voyage, she carried two 9-pound guns and four carronades. Special fittings included a raised upper deck. (From Dibner [1960], in the Burndy Collection at the Huntington Library, San Marino, Calif. Used with permission of the Huntington Library.)

six feet tall, a lean, sinewy, powerful, and inexhaustible young man.

For the next four months Captain FitzRoy refitted the ship, and Darwin packed and repacked the tiny poop cabin with books, microscope, specimen jars, etc. The *Beagle* had already made its first surveying voyage to South America (1826–30) and had returned with FitzRoy in command after the captain committed suicide in 1828. Charles was twenty-two and FitzRoy was twenty-six years old when the second voyage sailed off to make history. The *Beagle*'s mission was to complete the survey of the South American coast begun by the first voyage, take accurate longitude readings with the twenty-four chronometers aboard, and make geophysical measurements.

The *Beagle* sailed on 27 December 1831, after being forced back to port twice by storms. Darwin suffered terribly from seasickness for the next five years, often eating only raisins and dry biscuits and resting for hours in his cabin. Eventually he began to spend longer periods on shore when the *Beagle* docked at various ports. In South America, he would rent or borrow horses and ride inland, arranging to meet the *Beagle* at a certain time and place. A practiced hunter, Darwin often brought back fresh game for the ship's galley. On one inland trip lasting four months, he rode with gauchos, camped with Spanish bandits, landed in the middle of revolutions, and had to cope with soldiers, Indians, and whizzing bullets. The extraordinary experiences in the rainforests of Brazil, the arid pampas of Argentina, along the coast of Chile, and high in the Andes shaped the rest of his life and the history of science.

Darwin was filled with delight at the thought that he might

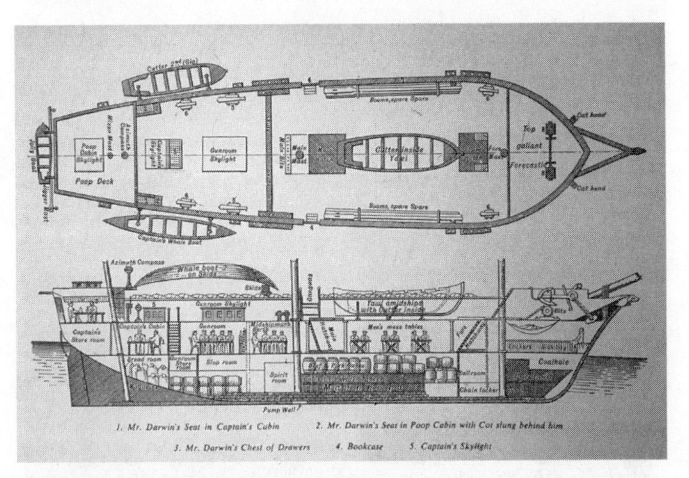

HMS *Beagle*, main deck and side elevation. Darwin shared the poop cabin (no. 2 on label) with draughtsman John Lort Stokes and midshipman Philip Gidley King, who made the drawing. (National Maritime Museum)

write a book on the geology of the places he visited. He was inspired by Humboldt's descriptions and greatly stimulated by another book he had with him, Charles Lyell's *The Principles of Geology*. Lyell rejected the catastrophist interpretation of the formation of the earth and developed the concept of uniformitarianism, which stated that the physical processes of sedimentation, erosion, and volcanic activity occurred in the

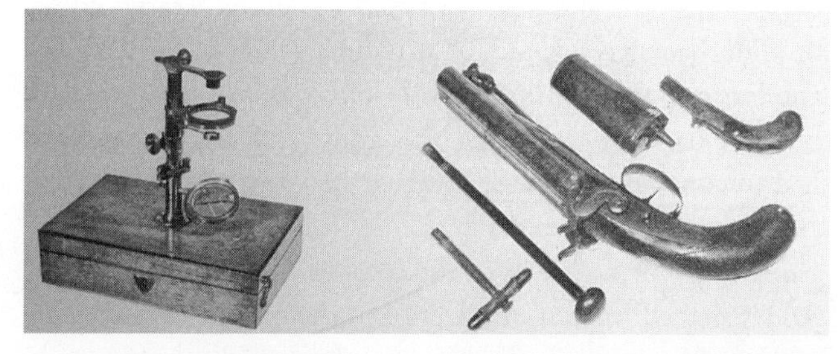

The microscope and pistols that Darwin had with him on the voyage of the *Beagle*. (Down House)

The tropics and its stunning biodiversity had a great impact on Darwin. A bird-hunter's view of a Brazilian forest. J. B. von Spix and C. F. P. von Martius, *Atlas zur Reise Brasilien in den Jahren 1817 bis 1829 gemacht* (1823–31). (Cambridge University Library)

Charles Lyell, author of *Principles of Geology* (3 vols., 1830–33) and one of Darwin's confidants. He was a Fellow of the Royal Society, and the most distinguished geologist of the day. Darwin wrote to J. D. Hooker that he felt that his own geology books came "half out of Lyell's brain." (British Museum)

past at about the same rate and frequency as they currently did. Darwin learned a great deal from Lyell's book, and they eventually became close friends, with Darwin later using the uniformitarian concept to make sense of the geological structures he was to describe.

Throughout the five-year voyage, Darwin recorded his observations in his journal. He sent many letters home to

Syms Covington, Darwin's personal servant on the *Beagle*, who remained in Charles's employ as secretary/servant until 1839. (Photographed in Australia. Ms. Marjorie B. Sirl of Bega, New South Wales, Australia.)

family and friends, including Professor Henslow. Charles also received mail when in port and was sad when he learned from his sisters that Fanny Owen was soon to be married.

In addition, Darwin assembled a huge collection of specimens. Syms Covington, "fiddler and boy to the poop cabin," was assigned to him as an assistant to help skin and clean birds and mammals and to sort and pack the shells, plants, bones, rocks, fossils, etc., which he shipped back to Henslow from various ports of call: Montevideo, Uruguay; Rio de la Plata, Argentina; the Falkland Islands; Buenos Aires, Argentina; and Valparaíso, Chile. Before he later moved to New South Wales, Australia, Covington was a personal servant, and then secretary/servant, to Darwin from 1833–39.

During the early part of the voyage in September 1832, about 400 miles south of Buenos Aires, Darwin spotted some bones and shells in the rocks of a low cliff. Using pickaxes,

Darwin and Covington uncovered huge fossil bones, including a massive jaw with a tooth. This was the skeleton of a giant sloth, *Megatherium*, and one of Darwin's most exciting discoveries. Only a single complete specimen was known in Europe, and extinction was an idea almost as revolutionary then as "transmutation," the word used for "evolution" at the time. FitzRoy's explanation for extinction was "the door of the Ark being made too small."

FitzRoy, a Tory aristocrat, and Darwin, a Whig gentleman, were only four years apart in age, but they held widely divergent views on religion, politics, and slavery. For the most part, they maintained the mutual respect necessary for two men living in such close quarters for an extended period. Yet Darwin was unaccustomed to Navy discipline and was upset by the severity of flogging, the standard shipboard punishment meted out to unruly crew members. For his part, FitzRoy fumed about the piles of "rubbish" Darwin collected. The crew was fond of Darwin because of his boundless energy on land and called him "Philos" as the philosophical member of the ship's company. But that did not spare him from being given the same hazing as any sailor crossing the equator for the first time.

An incident in Brazil is typical of the difficulty of remaining on good terms with the captain of a man-of-war. A slave was brought before a plantation owner and asked if he was happy with his life; the slave replied that he was. FitzRoy said this showed that slavery was not bad for the slaves, while Darwin ridiculed this position and said that the response of a slave in the presence of his master was not to be accepted at face value. FitzRoy flew into a rage and banned Darwin from his cabin. The gunroom officers invited Darwin to mess with them, and in a few hours FitzRoy apologized to Darwin and invited him back into the captain's cabin, where they usually took meals together.

Crossing the Line by August Earle, the *Beagle*'s first artist (on the second voyage), shows Darwin receiving the traditional lathering and dousing of a sailor who crosses the equator for the first time. (From FitzRoy's *Narrative of the Surveying Voyages of HMS* Adventure *and* Beagle *between the Years 1826 and 1836*)

Darwin, as the unpaid gentleman companion to the captain, was the only person on the ship who could be on intimate terms with FitzRoy and was FitzRoy's only outlet for friendship. Darwin was a stabilizing influence on this important and accomplished man who had a depressive and mercurial temperament. FitzRoy eventually became a member of parliament and then governor-general of New Zealand, and invented weather forecasting, before committing suicide nearly thirty years after the *Beagle* returned home.

The *Beagle*'s surgeon, Robert McCormick, also a naturalist, became envious of Darwin's position, quarreled with Captain FitzRoy and First Lieutenant John Clements Wickham (Darwin's favorite officer), and left the ship in Rio in 1832. The assistant surgeon, Benjamin Bynoe, became the acting surgeon for the remainder of the voyage and took care of Darwin during his illness at Valparaíso. Wickham commanded the third voyage of the *Beagle* (1837–43) and became the first Government Resident at Moreton Bay (Brisbane), Queensland, serving from 1853 to 1860. Another officer who befriended Darwin aboard the *Beagle* and often accompanied him on land expeditions was Second Lieutenant Bartholemew James Sulivan, who, like FitzRoy, eventually became an admiral in the Royal Navy.

The *Beagle*'s artist was Augustus Earle, but he left the ship in August 1832 because of illness. He was replaced by Conrad Martens, who joined the *Beagle* at Montevideo and later became a distinguished landscape painter in Australia.

Darwin was not the only passenger on the *Beagle*; there were four others. On the first voyage of the *Beagle*, FitzRoy had kidnapped four Fuegians from Tierra del Fuego and

Slavedriver punishing a slave, from *Brésil, Colombie et Guyane*. Drawing by Ferdinand Denis. (The British Library)

Repairing planks and copper sheathing on the hull of HMS *Beagle*, on the banks of the mouth of the Santa Cruz River in Patagonia; by Conrad Martens. *Inset*, Self-portrait drawn by Conrad Martens, the artist who replaced August Earle. (From *Narrative of the Surveying Voyages of HMS* Adventure *and* Beagle *between the Years 1826 and 1836*; and Mitchell Library, Sydney.)

taken them to England to educate and civilize them, with the thought of returning them to spread Christianity among their own people. These civilized natives should then be willing to help shipwrecked sailors who might drift ashore at some later time. FitzRoy named the woman Fuegia Basket and two of the men York Minster and Jemmy Button (whom he bought for a button). The fourth Fuegian, named Boat Memory, died in England from an overdose of smallpox vaccine.

Richard Matthews, of the Church Missionary Society, was traveling with the three surviving Feugians. This party was left to join the Feugians' tribe at Tierra del Fuego and provided with some supplies. When the *Beagle* called again nine days later, the natives had reverted to their tribal ways, and

The three surviving Fuegians that FitzRoy returned to their homeland on the second *Beagle* voyage. Fuegia Basket (*top*), Jemmy Button (*middle*), and York Minster (*bottom*), drawn by Captain FitzRoy in 1833. FitzRoy tried to "civilize" the Fuegians by taking them to England on the first *Beagle* voyage. (From *Narrative of the Surveying Voyages of HMS* Adventure *and* Beagle *between the Years 1826 and 1836*)

Conrad Martens' drawing of a man of the Yapoo Tekeenica tribe and his dog at Portrait Cove, Tierra del Fuego. (From *Narrative of the Surveying Voyages of HMS* Adventure *and* Beagle *between the Years 1826 and 1836*)

Matthews had to be rescued. The *Beagle* called again one year later, and Jemmy Button rowed out to the ship naked. He told FitzRoy that York Minster and Fuegia Basket had stolen his possessions and moved to a different island. FitzRoy's experiment had failed, and he was greatly depressed.

On 29 January 1833, Darwin and a landing party were on

a beach in the Beagle Channel at Tierra del Fuego, admiring the "beryl-like blue" glaciers that extended from the mountains to the shore and the icebergs formed by fragmentation from the glaciers. The *Beagle*'s boats were on shore when a huge chunk of ice calved from the glacier and generated an enormous wave that came roaring at Darwin and the men. In *Voyage of the Beagle*, Darwin wrote that "the men ran down as quickly as they could to the boats; for the chance of their being dashed to pieces was evident . . . and the boats, though thrice lifted on high and let fall again, received no damage. This was most fortunate for us, for we were a hundred miles distant from the ship."

What Darwin failed to mention, with his usual modesty, was that he was one of the men who saved the boats. FitzRoy described the event in his journal: "By the exertions of those who grappled them [the boats] or seized their ropes, they were hauled up again out of reach of a second and third roller; and indeed we had good reason to rejoice that they were just saved in time; for had not Mr. Darwin, and two or three of the men, run to them instantly, they would have been swept away from us irrevocably." FitzRoy named Darwin Sound and Mt. Darwin in this region for his friend, another thing Darwin neglected to mention.

FOUR

Discovery

DURING AN EXPEDITION from Valparaíso to the high Andes in 1834, Darwin located marine deposits that included the remains of a petrified forest. He deduced that this part of the South American continent had once been under the sea and was subsequently uplifted more than 7000 feet.

Darwin was on Chiloe Island when he witnessed the volcanic eruption of Chile's Mt. Osorno on 26 November 1834.

He also survived an earthquake in Valdivia which destroyed the city of Concepción on 20 February 1835. These experiences demonstrated Lyell's geological principles in dramatic fashion and stimulated Darwin's revolutionary ideas.

During an expedition into Argentina, and again in the Andes, Darwin fell seriously ill. He developed a fever, now suspected to have been caused by the bite of the benchuga bug, *Triatoma infestans*, a frequent carrier of the protozoan parasite *Trypanosoma cruzi*, which causes Chagas's disease. *Trypanosoma* frequently invades cardiac muscle and destroys nerves in the intestines, causing heart and digestive disorders; it also has periods of latency. Today approximately 50,000 people in Latin America die annually from Chagas's disease, which kills 10 to 20 percent of the people it infects.

This disease has been offered as one explanation of Darwin's lifelong illness, but not all medical detectives are convinced. Others consider his physical problems, although very real, to have been psychosomatically induced, caused by the stress his

Traveling the Andes, which provided Darwin with evidence of the geological forces described by Lyell in his *Principles of Geology.* (From Darwin's *Journal of Researches*)

ideas produced, and there is a good deal of circumstantial support for this hypothesis. In fact, Darwin's health improved in the years after the general scientific acceptance of his ideas. His stress-based symptoms might even be considered a form of panic disorder. Still others have speculated that he turned himself into an invalid to avoid social distractions so he could get on with his work. These three explanations are not mutually exclusive. Darwin may very well have had a mild form of Chagas's disease as well as suffering the ill effects of self-induced stress, and he certainly guarded his time. (See *To Be an Invalid* by Ralph Colp Jr. for speculation about Darwin's illness.)

The eruption of Antuco volcano. Witnessing volcanic eruptions helped create Darwin the geologist. (From *Atlas de la Historia Fisica y Politica de Chile*, by C. Gay [1854].)

Remains of the cathedral at Concepción, Chile, after the earthquake in 1835, from a drawing by Darwin's favorite *Beagle* officer, First Lieutenant John Clements Wickham. Earthquakes, volcanoes, and mountain-building completed Darwin's geological education. (From *Narrative of the Surveying Voyages of HMS* Adventure *and* Beagle *between the Years 1826 and 1836*.)

After departing from the west coast of South America, the *Beagle* sailed west about 600 miles (1000 km) to the Galápagos Islands, named for their giant tortoises. The *Beagle* spent five weeks in the Galápagos in 1835, and Darwin visited four of the sixteen major islands. During a dinner party, the vice-governor of the Galápagos explained that he could identify which island a tortoise came from by the shape of its shell. (Later, when leaving the Galápagos, the crew took eighteen live tortoises to use as food on the way home.)

Likewise, Darwin had noted differences between the mockingbirds (four species of *Nesomimus*) from various islands. Almost every biology textbook has a pictorial feature about Darwin's finches and how important they were to Darwin's idea of natural selection. The truth is that Darwin, while in the Galápagos, did not recognize differences among the finches. In fact, his labels did not include the island from which he collected each finch. Only after the voyage, when ornithologist

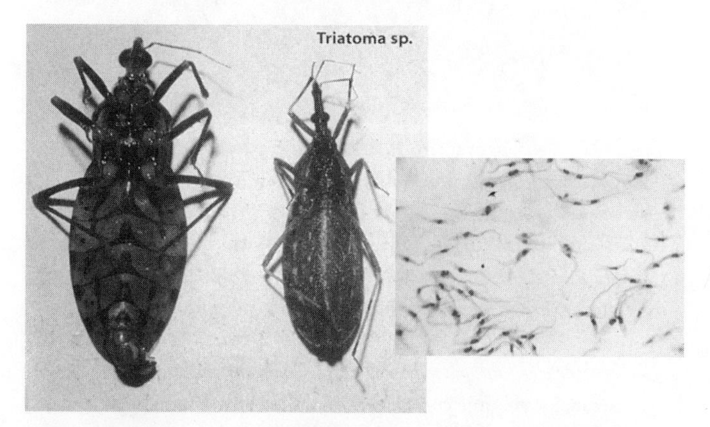

Triatoma sp., the vector of Chagas's disease caused by the protozoan parasite *Trypanosoma cruzi* (*inset*). These assassin bugs are members of the order Hemiptera, family Reduviidae. (From *Missouri Conservationist,* June 2003, p. 27; used with permission of the photographer, David Liebman.)

Top, Galápagos giant tortoise, *Geochelone elephantopus porteri*, from Santa Cruz. Note the dome-shaped shell found on tortoises from islands with rainfall adequate to support lush vegetation. These tortoises tend to be larger and have relatively short legs and necks. These adaptations allow the tortoise to push its way through vegetation and graze on grasses. There are fourteen races or subspecies of Galápagos tortoises. (Author's photograph) *Bottom*, Galápagos giant tortoise, *Geochelone elephantopus* subsp., from Isabela, the largest island. Isabela has six major volcanic cones, and five are occupied by different subspecies of tortoise. These populations are isolated from each other by extremely arid, barren, and rugged lava flows. Note that the front of the carapace is elevated. This allows the neck to be greatly extended, which facilitates browsing on cactus or "trees." Smaller tortoises with saddleback-shaped carapaces are found on Española and Pinta islands. (Author's photograph)

and artist John Gould began to study and illustrate the finches, did it become clear that they, like the mockingbirds, were different from island to island.

Darwin had to rely on FitzRoy's notes and specimens to determine where the finches were collected. Although he did not fully appreciate all the evolutionary implications of what he saw until nearly eighteen months after his return from the Galápagos, its fauna, including the marine and terrestrial iguanas, provided him with some strong insights into transmutation (evolution). Darwin considered his *Beagle* experiences to be "the first real training or education of my mind."

The *Beagle*, having finished its survey, headed south across the Pacific for Tahiti, New Zealand, and Australia—calling in at Sydney, Hobart, and Georges Sound (Western Australia)—and then into the Indian Ocean.

John Lort Stokes was a nineteen-year-old mate and assistant surveyor when the second voyage of the *Beagle* left England. He shared the poop cabin with Darwin and midshipman Philip Gidley King, who remained in Sydney after the *Beagle* docked there in 1836. On the third voyage of the *Beagle*, in 1839, Stokes, promoted to naval surveyor, named a beautiful harbor after Charles Darwin, his friend and former shipmate, to let him know he was not forgotten. Darwin Harbour (later to become the City of Darwin, capital of Australia's Northern Territory) was named in Charles's honor, but he never visited the area. By the end of the third voyage, Stokes was in command of the *Beagle*, and Stokes Hill Wharf in Port Darwin is named for him.

While in the Cocos (Keeling) Islands, a cluster of twenty-seven small coral islands 1700 miles (2800 km) northwest of Perth, Darwin verified his ideas about coral reef formation. He suggested that reefs originally became established on the shores of recent volcanic islands through the colonization of larvae from nearby reefs. As an island gradually subsided under its own weight and the weight of the expanding coral growth, the

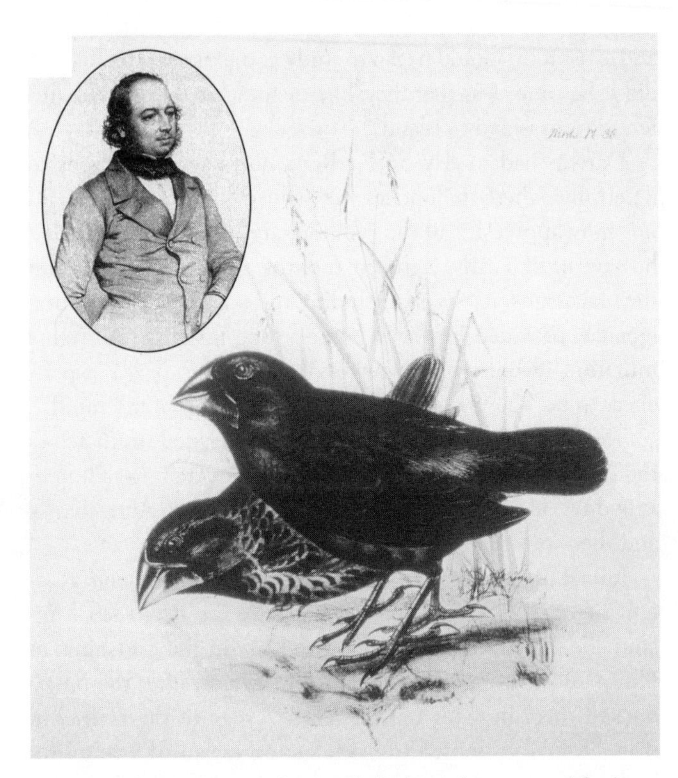

Geospiza magnirostris, one of Darwin's finches, a large ground finch with a bill adapted for cracking large seeds. Painted by John Gould (*inset*) and published in the *Zoology of the Voyage of H.M.S.* Beagle. Darwin's finches are the classic example of adaptive radiation, whereby one species gives rise to multiple species that occupy different niches. The fourteen species from the Galápagos Islands can be divided into tree finches, which are primarily insect eaters, and ground finches, which mostly feed on seeds. Beak size reflects the size of seeds or insects that are eaten. In *The Voyage of the Beagle*, Darwin wrote that "seeing this gradation and diversity of structure in one small, intimately related group of birds, one might really fancy that from an original paucity of birds in this archipelago, one species had been taken and modified for different ends." (Finches from *Zoology of the Voyage of H.M.S.* Beagle [part 3, plate 36]. Gould portrait at age forty-nine by T. H. Maguire, in 1849; British Museum.)

fringing corals along its shores grew upward. This upward growth more or less matched the rate of subsidence of the island.

Darwin's theory of coral growth has been confirmed by drilling on atolls, which revealed thick layers of reef material over volcanic rock. On Eniwetok Atoll, a drill bored through a coral layer over 1000 meters (3,281 feet) thick before basalt was encountered. Darwin's theory of coral growth is more than most of us can expect to accomplish in a lifetime, but it was just one of his many contributions to biology.

The *Beagle* sailed for home by way of the Cape of Good Hope, St. Helena, and Ascension Island. At Ascension Island, Charles received a letter from his sister Catherine informing him that Henslow had circulated his letters and specimens and that Charles was famous. This letter left Charles proud, puzzled, and embarrassed.

In order to formally complete a circumnavigation of the globe, FitzRoy sailed the *Beagle* back to the coast of South America, then north to home. The *Beagle* landed at Falmouth, England, on 2 October 1836, four years, nine months, and five days after it sailed.

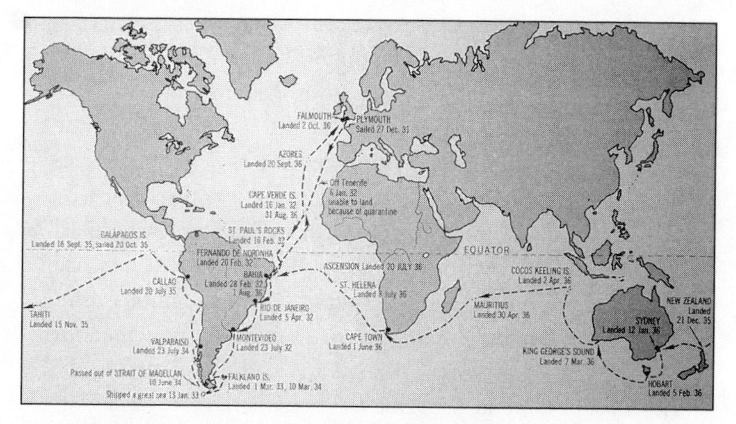

Map of the second voyage of the *Beagle*, from 27 December 1831 to 2 October 1836, under the command of Captain FitzRoy. Total time away from England was 1,737 days. (Publisher's Art, Inc., from Gregor [1967].)

FIVE

Maturity

UPON LANDING IN FALMOUTH on 2 October, Darwin made a beeline for home, where he was greeted by his father and three sisters. Ten days later, he was back at Cambridge with Professor Henslow. He was now a confident young man with a new direction in life. He no longer wanted to be a clergyman, having earned the approval of the scientific community and, most importantly, of his father. Henslow had shown Darwin's specimens and descriptive letters to the Cambridge Philosophical Society, and Darwin was well known and admired among scientists because of his work during the *Beagle* voyage.

When Darwin moved to London fifteen days after arriving back in England, he was astonished to learn that his *Megatherium* had been displayed at the 1833 meeting of the British Association for the Advancement of Science and had generated a great deal of excitement. He was befriended by some of London's most notable scientists and was introduced to Charles Lyell, who was then president of the Geological Society. Lyell subsequently introduced him to famous comparative anatomist Richard Owen. Darwin was presented for membership in the Geological Society and was elected to the Athenaeum Club at the same time as Charles Dickens. Darwin's presentation to the Geological Society (with Lyell in the audience) on 4 January 1837, on the elevation of the coast of Chile, made a significant impression.

Erasmus Alvey Darwin, Charles's elder and only brother, by George Richmond, in 1841. Erasmus trained as a physician, but never practiced. He lived the life of a London socialite, off his allowance and inheritance from Robert Darwin. (Darwin Museum, Downe)

Darwin sent back 1,529 species bottled in alcohol and 3,907 dried specimens, and he sought out various taxonomic experts to deal with his collections of fishes, reptiles, birds, mammals, and fossils. He kept the rocks, invertebrates, and plants to describe himself. He moved to Cambridge for three months to deal with the specimens and work on the *Journal of Researches*, based on his daily journal during the voyage, and later completed the *Journal* in London. This book was first published as volume 3 of a narrative edited by FitzRoy, and then in its own right in 1839, as the *Journal of Researches into the Geology and Natural History of Various Countries Visited by*

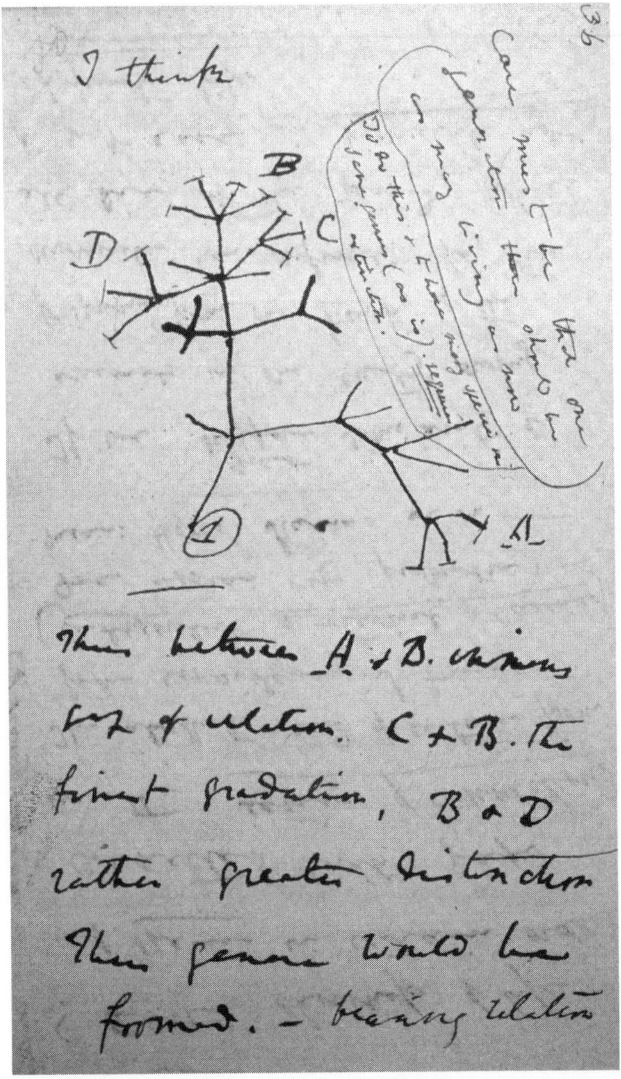

Darwin's "tree of life" from his transmutation notebook B, page 36 (1837). This was the first evolutionary tree ever sketched by anyone and is Darwin's first attempt to show the relatedness of all animal life. (Cambridge University Library)

H. M. S. Beagle *Round the World.* It was Darwin's first book and is now universally known as *The Voyage of the Beagle. The Zoology of the Voyage of the Beagle,* edited by Darwin, was published in five parts, from 1838 to 1843, and Darwin continued to work on *The Geology of the Beagle.*

In London, Darwin resided in Great Marlborough Street to be close to his socialite brother, Erasmus. Although he earned a medical degree from Edinburgh, Erasmus basically lived off his inheritance from Dr. Darwin. Charles also socialized with Lyell, often patronized the Athenaeum Club, and was a guest in the home of Harriet Martineau, feminist, social reformer, and author, who was an intimate friend of Erasmus. Darwin's social life was not the only thing that had altered since he set sail on the *Beagle*. During his absence, Britain had changed greatly with the passage of the Reform Bill of 1832, which gave real political power to the middle class.

In July 1837, Darwin secretly entered his ideas about the transmutation of species (the word "evolution" was not commonly employed in the way we use it today), although he had not yet thought of the concept of natural selection. He was to fill four "transmutation notebooks." In notebook B, he drew an irregularly branching tree to represent the common ancestry of all animals, a famous sketch that is the first representation of an evolutionary tree. At the top of the page, he penned, "I think." At this time he also noticed the heart palpitations he was to experience for the rest of his life, especially during periods of stress—such as when he considered his heretical evolutionary ideas.

SIX

A Proposal

DARWIN LONGED FOR a more quiet life in the country, to escape from polluted, crowded London. His thoughts also turned to marriage. In keeping with his rational nature, he listed the advantages and disadvantages of marriage, and somehow the odds came down in favor of it.

Marry
Children—(if it Please God)—Constant companion, (& friend in old age) who will feel interested in one,—object to be beloved & played with.—better than a dog anyhow.—Home, & someone to take care of house—Charms of music & female chit-chat.—These things good for one's health.—*but terrible loss of time.*—My God, it is intolerable to think of spending ones whole life, like a neuter bee, working, working, & nothing after all.—No, no won't do it.—Imagine living all one's day solitarily in smoky dirty London House.—Only picture to yourself a nice soft wife on a sofa with good fire, & books music perhaps—Compare this vision with the dingy reality of Grt. Marlbro' St. Marry—Marry—Marry Q.E.D. [*quod erat demonstrandum* = "which was to be demonstrated"]

Not Marry
Freedom to go where one liked—choice of Society & *little of it.*—Conversation of clever men at clubs—not forced to visit

relatives, & to bend in every trifle.—to have the expense & anxiety of children—perhaps quarelling—*Loss of time.*—cannot read in the Evenings—fatness & idleness—Anxiety & responsibility—less money for books &c—if many children forced to gain one's bread.—(But then it is very bad for ones health to work too much) Perhaps my wife wont like London; then the sentence is banishment & degradation into indolent, idle fool.

Emma Wedgwood (1808–1896) was his first cousin, the ninth and last child of Uncle Jos, and Charles's friend since childhood. He had often visited her and her siblings at Maer while he was growing up. She had all the qualities he desired in a wife, and both families welcomed his attention to Emma. With some paternal advice, Charles proposed, and Emma accepted on 11 November 1838 at Maer. On 24 January 1839, Charles was elected to the Royal Society of London, and on 29 January 1839, at St. Peter's Church in Maer, Charles and Emma were wed. They started married life as a wealthy couple, due to Emma's £5000 dowry from the Wedgwood fortune and Dr. Darwin's gift of £10,000 plus an allowance of £400 a year. Just two years earlier, Charles's sister Caroline had married Emma's brother Josiah Wedgwood III. Charles's mother was also a Wedgwood, so the Darwins and the Wedgwoods were firmly intertwined.

After accepting his marriage proposal, Emma wrote that Charles "is the most open transparent man I ever saw, and every word expresses his real thoughts. He is particularly affectionate and very nice to his father and sisters, and perfectly sweet-tempered." Emma was thirty and Charles was twenty-nine when they married. She was a spirited, educated woman who spoke French, Italian, and German. She played the piano regularly and, as a young woman, had once taken a few lessons from Chopin during an extended family tour to Europe. Emma dedicated her life to Charles and to the demands of his constant illness and work, and their initial friendship grew to a deep, mutual devotion.

Emma, nominally Anglican, was deeply religious in the Unitarian tradition, and Darwin's ideas were challenging to her beliefs. Nevertheless, she read his manuscripts before he sent them to scientific colleagues. They had a house in Upper Gower Street (London), called Macaw Cottage because of its gaudy curtains. On 27 December 1839, their first child, William Erasmus (1839–1914), was born. (William eventually supervised the family's financial affairs with great success.) This happy event was greeted with enthusiasm by Darwin the scientist. He recorded observations on the behavior of his "little animalcule of a son" (and also began noting his observations of an orangutan in the London Zoo). William Erasmus and all the subsequent Darwin children received much affection and attention from Charles, a characteristic that Emma found endearing and one that was completely unlike the image of a distant Victorian father.

In his autobiography, Darwin wrote: "In October 1838, that is fifteen months after I had begun my systematic enquiry, I happened to read for amusement Malthus *On Population*, and being well prepared to appreciate the struggle for existence which goes on from long-continued observation of the habits of animals and plants, it at once struck me that under these circumstances favorable variations would tend to be preserved, and unfavorable ones to be destroyed. The result of this would be the formation of new species. Here, then, I had at last got a theory by which to work."

From 1836 to 1844, Darwin filled fifteen notebooks with his ideas about geology, the transmutation of species, and metaphysical enquiries. These notes, which have been transcribed, became the outline for his future research and publications and show the gestation of his ideas.

In March 1841, the second Darwin child, Anne Elizabeth

Charles Darwin at thirty-three with his oldest child, William, in an 1842 daguerreotype. This is the only known photograph of Darwin with a family member. (Library, University College, London)

(1841–1851), was born, adding to the family joy. Darwin was eager to withdraw from the demands of society and was thinking of moving from London. In the summer of 1842, Darwin penned a brief 35-page, 1500-word sketch of his species theory. He felt confident about what he now referred to as a "natural means of selection." He also sent his *Coral Reefs* book (part of *The Geology of the Voyage*) to the publishers. By now Emma was expecting their third child, and Macaw Cottage was becoming too small for the growing family and their servants. Charles's father agreed to help finance the cost of a country house, and they made the decision to move.

SEVEN

Life at Down House

AFTER SEVERAL SEARCHES, the Darwins decided on a property near the small village of Downe in Kent, about sixteen miles south of London. The price was £2200. They moved in by mid-September 1842. Mary Eleanor, their third child, was born on 23 September, just a few days after the move, but died three weeks later, on 16 October, so the first few months at Down House were sad.

The name of the village is Downe, with an *e* that was added just before the Darwins arrived. However, they did not want to change the spelling of the house, which remains Down House. The village of about forty houses was home to tenant farmers and agricultural laborers. The pub, George & Dragon Inn, and church, St. Mary the Virgin, still stand in the center of the village. In 1842, Down House was a squarish Georgian building, set on fifteen acres. The Darwins added rooms as the number of family members and servants grew. A high flint and brick wall was built in the front to provide privacy, and the road was lowered so people in coaches could not see over the wall. Gardens and greenhouse were rebuilt. Darwin also purchased an adjacent three-acre strip of land that became his beloved Sandwalk. He kept a pile of stones by the trail and would kick one off each time he completed a circuit of the Sandwalk.

Darwin personally supervised all the alterations to the

Gates and brick and flint wall in front of Down House in 1974. Shortly after the Darwins purchased Down House, the road level was lowered to provide privacy from the prying eyes of people in carriages. (Author's photograph, 1974)

Darwin's Sandwalk. His "thinking path" meanders through shady woods and emerges along a sunny meadow. He was often accompanied on his daily walks by his fox terrier, Polly. (Author's photograph, 1974)

building and grounds, understood what price his hay could command, and eventually took an active part in parish and civic duties. He was even an honorary magistrate and heard cases in Petty Sessions Court. Emma conducted the Sunday school, taught reading to village children, and took food and medicine to poor families in the village.

Darwin wrote to FitzRoy that "my life goes on like clock-work, and I am fixed on the spot where I shall end it." He was home for good. He made briefs visits to friends and took family vacations, but his voyaging was over.

Erasmus Darwin, physician, poet, philosopher, and Charles Darwin's grandfather. (Portrait by Joseph Wright in 1770. National Portrait Gallery.)

Wedgwood family at Eturia Hall, painted by George Stubbs in 1780. Susannah Wedgwood (Charles Darwin's mother) is seated on the horse *left of center*. *Right to left*: Josiah Wedgwood I and wife Sarah seated on the bench around the tree, children John and Josiah II on horseback, Catherine, Thomas (on horse), Sarah, and Mary-Anne. *Inset,* Susannah (née Wedgwood) Darwin, Charles's mother, at age twenty-seven, in miniature portrait. (Wedgwood Museum; Darwin Museum, Down House)

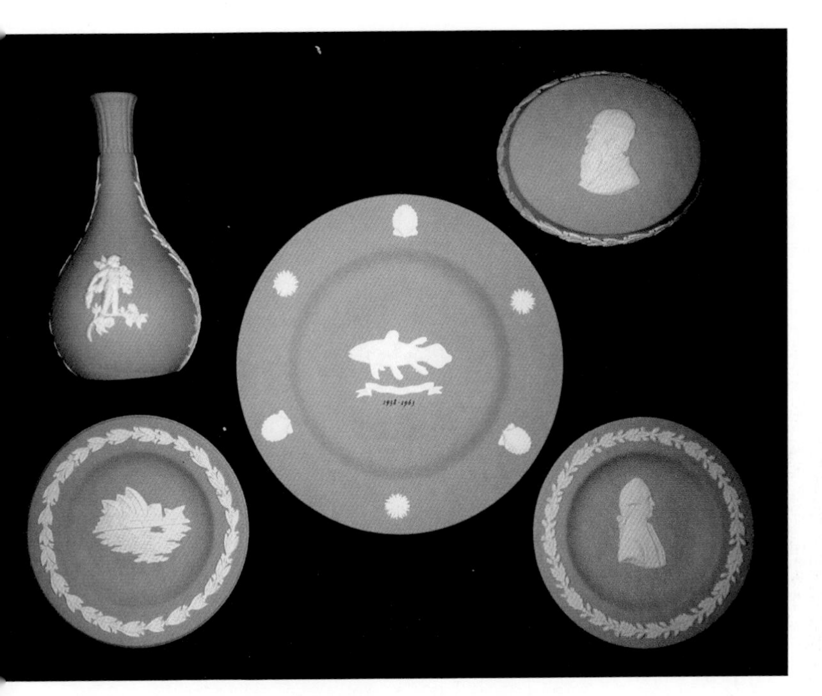

Wedgwood ware in the author's collection. *Clockwise from bottom left*: Sydney Opera House (dedicated 1973); bud vase; Coelacanth saucer (gift of Mrs. J. L. B. Smith to the author); Darwin jewelry box; Captain James Cook Australian Bicentennial, 1770–1970.

The earliest portrait of Charles (age seven) and sister Catherine (age six), in chalk, by Rolinda (?) Sharples, drawn in 1816. (Darwin Museum, Down House)

Robert FitzRoy, after his promotion to vice-admiral in 1863. He was governor-general of New Zealand (1843–45) and became chief of the meteorological department of the Board of Trade in 1854. He visited Down House in 1857, the last time Darwin and FitzRoy were together. (Oil painting by Francis Lane; Royal Naval College, Greenwich)

Midshipmen's berth by *Beagle* artist Augustus Earle. (National Maritime Museum, Greenwich)

HMS *Beagle* off James Island, Galápagos. (Painted by John Chancellor, 17 October 1835; Alexander Gallery.)

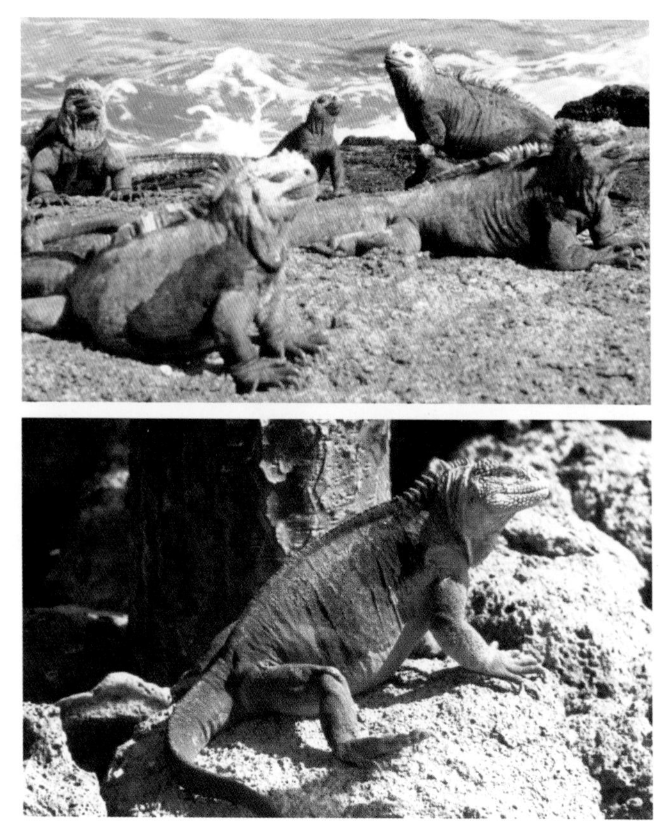

Top, Marine iguana, *Amblyrhynchus cristatus,* from Fernandina, Galápagos Islands. Their stubby snouts help them graze algae from submerged rocks. They can remain underwater for as long as an hour and dive to more than twelve meters (39 feet). On land they often bask in large colonies, and their snouts become coated with salt, which they expel from their nostrils by sneezing. (Author's photograph) *Bottom,* Galápagos land iguana, *Conolophus subcristatus*, from South Plaza. Land iguanas feed upon the fruits and pads of prickly pear cactus, *Opuntia* sp., removing the cactus spines by scraping them with their claws. These iguanas are found on the drier parts of the central and western islands and may grow to weigh six kilograms (13 lbs). (Author's photograph)

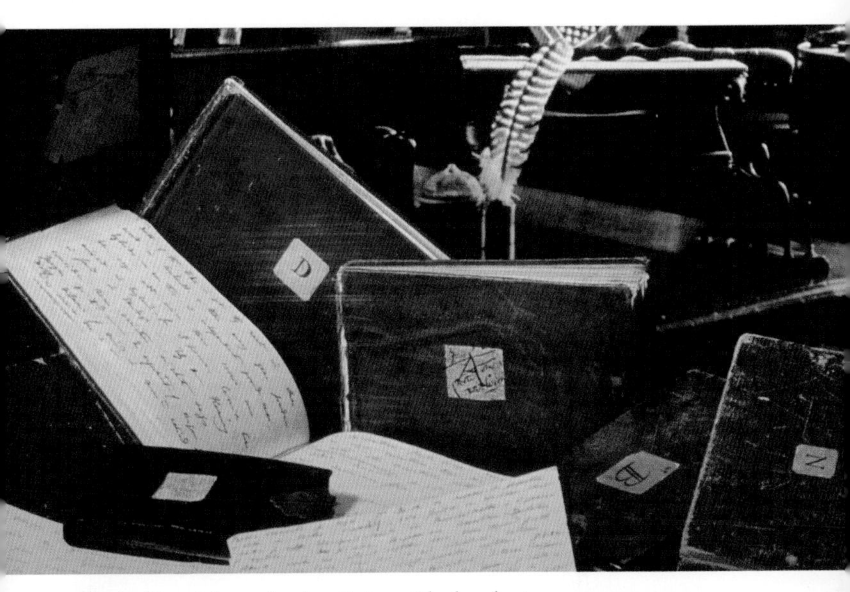

Charles Darwin's notebooks, 1836–44. The handwritten contents, transcribed by Barrett et al. (1987), reveal Darwin's thought processes and provided the outline for several of his books.

Wedding portraits of (*above*) Charles (watercolor) and (*left*) Emma (chalk) Darwin in 1840, at age thirty and thirty-one, respectively. (By George Richmond; Down House)

Front of Down House, summer 1974. (Author's photograph)

The drawing room at Down House, with Emma's piano and a couch for Charles to lie on while Emma played. (Author's photograph, 1974)

Rear of Down House, summer 1974. The vine-covered windows in the lower right illuminate Charles Darwin's study. (Author's photograph)

Darwin's study at Down House. He wrote *The Origin of Species* and other works on a cloth-covered board that rested on the arms of his chair. The iron-framed chair had wheels, so he could glide around the room with minimal effort. Multi-shelf files and a chest of drawers are behind the chair. His square work table is in front of the writing chair, and the round specimen table is to the right. Portraits over the fireplace are, *left to right*: Joseph Hooker, Charles Lyell, and Josiah Wedgwood, Darwin's maternal grandfather. Reflected in the mirror are parts of Darwin's library. (Author's photograph, 1974)

Color plate from Darwin's barnacle monograph. His barnacle research consumed eight years (1846–54).

Thirty-two fancy pigeon breeds produced from the common rock dove, *Columba livia*, by artificial selection. This helped Darwin visualize what nature could do given millions of years of natural selection. (Painting by British artist A. F. Lydon from *Boy's Own Paper*.)

EIGHT

Correspondence

DARWIN HAD INTENDED to get to London for a night or two each month, so he would not become a "Kentish hog." But the eight-mile carriage ride to the railroad station was too much for his ill health. He reveled in the isolation of Down. He spent each day working in his study and enjoyed family activities as a restful diversion. He carried out his

Joseph Dalton Hooker, director of the Royal Botanic Gardens, Kew. He was Darwin's closest friend, confidant, and correspondent and a Fellow of the Royal Society. (British Museum)

communications with scientists via letters—lots of them. The Darwin Correspondence Project (see appendix C) has found about 14,500 letters to and from Darwin and 2000 of his correspondents; Darwin himself wrote about 7000 of them. He kept up a lifelong correspondence with Charles Lyell, and Joseph Dalton Hooker, director of the Royal Botanic Gardens (Kew Gardens), was another long-time friend and advisor with whom Darwin exchanged about 1400 letters. Zoologist Thomas Henry Huxley was a third favorite correspondent, who later became known as "Darwin's Bulldog" for his staunch defense of Darwin's ideas.

Darwin's inner circle thus included the greatest geologist, botanist, and zoologist of the day, and we can trace the development of his ideas from these letters. In 1844 he wrote to Hooker that "I am almost convinced (quite contrary to the opinion I started with) that species are not (it is like confessing a murder) immutable."

NINE

Daily Routine

DARWIN'S DAILY ROUTINE has been expertly chronicled by Hedley Atkins (1973) and summarized in Louise Wilson and Solene Morris's text for *Down House* (2000), from which the following is taken.

Darwin would rise fairly early and go for a walk before breakfast. From about 8:00 a.m. he would work in his study for an hour and a half, then he would take a break and listen to Emma reading family letters.

He would return to work until mid-day and then take his daily stroll around the Sandwalk, rain or shine. He was usually accompanied by his fox terrier, Polly. (At various times the family dogs were named Bobby, Button, Dicky, Pepper, Polly, Quiz, Tony, and Tyke.)

The main meal of the day was lunch, served at about one o'clock. He read the newspaper and then wrote letters or read until 3:00 p.m., when he would rest, often listening to Emma read a novel.

At about 4:30 p.m. he resumed work until 5:30, and then rested. A simple supper was served at 7:30, followed by a couple of games of backgammon with Emma or listening to her play the piano or read. Ever the data collector, Charles recorded each game and wrote to his botanist friend Asa Gray that "she poor creature has won only 2,490 games, whilst I have won, hurrah, hurrah, 2,795 games!"

Darwin's children remembered him as very patient and kind

Library in Darwin's study, directly across from his chair, with a work table in between. Many of the volumes on display belonged to Darwin and are on loan from Cambridge University, where his son Francis bequeathed them. (Author's photograph, 1974)

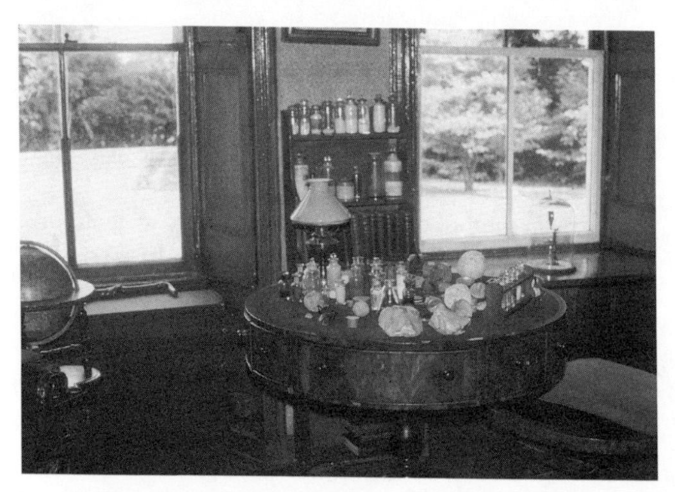

Round table in Darwin's study, a revolving drum that facilitates access to various specimens and chemicals on the table like a lazy Susan. (Author's photograph, 1974)

The screened area next to the fireplace is the lavatory. This was a convenience necessitated by Darwin's frequent ill health. The armchair belonged to Darwin's father. (Author's photograph, 1974)

Dining room at Down House. The dining table is now used to display documents. Over the mantle is a portrait of Darwin by John Collier, painted in 1881. To the right is a portrait of Emma. (Author's photograph, 1974)

Emma playing the piano for Charles in the drawing room of Down House. (Royal College of Surgeons, Down House)

in spite of his constant work and illness. He never scolded the children for interrupting him, even when they barged into his study. He took great personal interest in them and played with them whenever he could. The Darwins' fourth child, Henrietta Emma (1843–1929), was born on 25 September 1843. She eventually became Mrs. Richard Litchfield, the only Darwin daughter to marry. In *Period Piece* (1971), granddaughter and artist Gwen Raverat, daughter of Darwin's second son, George Howard (1845–1912)—a mathematician and Fellow of the Royal Society who was knighted in 1905—described many happy memories of the Darwin household.

In 1844, Darwin revised and augmented the brief 1842 sketch of his evolutionary theory, expanding it to 230 pages. He showed this 52,000-word essay to J. D. Hooker, but no one else. Also in 1844, an anonymously published book, entitled *Vestiges of the Natural History of the Creation*, caused a scandal in respectable society. It was an epic that told the story of Earth and life on it from its beginnings. The author, later revealed to be the journalist and publisher Robert Chambers, suggested the perpetual transformation of species in no uncertain terms.

The book was immensely popular, exciting those people longing for social reform. Darwin felt it was admirably written but full of geological and zoological errors. However, the uproar it created was undoubtedly one of the reasons why Charles delayed his own publication on evolution for another fifteen years. Darwin anticipated such a reaction to an account of a godless origin for species, and he wanted more time to develop his evidence. Aware of his failing health, he gathered up the enlarged manuscript of 1844 and added an extraordinary letter to Emma, including the specification of £400 to cover the cost of publication in the event of his sudden death.

Caricature of Darwin's children and friends sliding down the banister at Down House, causing Charles to emerge from his study. (From John Leech's 1850 etching, *Young Troublesome*.)

TEN

Taxonomy and Selection

THE SOLUTION TO THE VEXING QUESTION of why descendents diverged from their ancestors came to Darwin one day while he was riding in his carriage near home. He wrote in his autobiography that "the solution . . . is that the modified offspring . . . tend to become adapted to many and highly diversified places in the economy of nature." This ecological niche concept helped him formulate the crux of his argument—that natural selection picks out the best-adapted variation for a given environment.

Hooker wrote to Darwin that "no one has hardly a right to examine the question of species who has not minutely described many." So with his *Geological Observations on South America* ready for publication in October 1846, Darwin took Hooker's advice to heart and turned his attention to the Cirripedia—the barnacles. This was his contribution to formal taxonomy, and these studies lasted eight years, dealing with 10,000 specimens of all known living and fossil forms of barnacles. This taught him to think about homology (the structural similarity between parts of different organisms) and embryology and made him a respected systematist.

Midway through his research, in frustration at its slow pace, he wrote, "I hate a barnacle as no man ever did before." However, he persevered and produced a monumental four-volume work that is still held in high esteem. From the barnacles, Darwin learned about the wide range of variation a

Darwin in 1849, at age forty, by T. H. Maguire. He was already a Fellow of the Royal Society. (The Bettman Archives)

species could possess and how each species was adapted to its environment. He also deduced how crablike ancestors gave rise to barnacles. His family became so accustomed to his spending hours in his study minutely examining barnacles that one of his boys once asked a friend, "Where does your father do his barnacles?"

Daughter Elizabeth "Bessy" (1847–1928) entered the Darwin household on 8 July 1847, and son Francis (1848–1925) was born on 16 August 1848. He later became a botanist, qualified as a physician, was elected a Fellow of the Royal Society, and was knighted in 1913.

Darwin's barnacle work was interrupted by his father's death in November 1848 as well as by the further decline of his own health. His stomach problems were so severe that some days he vomited constantly and could barely work. A friend recommended the trendy "water cure," and Darwin took his entire family to Dr. James Gully's hydropathic facility at Malvern, near the border of Wales. The spa treatments—which many,

Dr. Gully's Hydropathic Establishment at Malvern. *Inset*, Dr. James M. Gully, physician and owner of the spa at Malvern, who treated Charles with cold water and made him give up the use of snuff. (Gerald Morice Collection; Wellcome Trust)

even then, recognized as quackery—were based on the ability of cold water to stimulate circulation; Darwin was wrapped in wet sheets and doused with cold showers.

He planned to be away for only six weeks, but the whole family enjoyed the holiday and stayed for four months. Either the treatment or, more likely, the escape from the stress of his work and ideas made him feel much better. From this time onward, Darwin took a cold shower and morning scrub in the garden and believed that this improved his health.

The Darwins' eighth child, Leonard (1850–1943), was born on 15 January 1850. He eventually became a major in the Royal Engineers.

In the summer of 1850, the Darwins' eldest daughter Annie, already weakened by scarlet fever, started to show signs of illness. Her condition deteriorated, and in late March 1851 Charles took Annie, her sister Henrietta, their nurse, and their governess to Malvern. Emma was eight months pregnant, and remained at Down with the other children. In mid-April, at the age of ten, Annie died of tuberculosis, which was known as consumption in those days.

Charles was especially close to Annie, and it was a loss from which he never recovered. Her death was also an important factor in Darwin's abandoning his last vestiges of Christianity. Charles could not imagine a just and merciful God who would allow such suffering in innocent children. (Randal Keynes, Darwin's great-great-grandson, described the touching story of Charles, Emma, and Annie during this time in his book *Annie's Box* [2001].)

Emma Darwin at forty-five, with son Leonard (age two and a half) in 1853. Emma was the ninth and last child of Josiah Wedgwood II. The children called her Mammy, and Charles referred to her as "my greatest blessing." He also wrote: "I marvel at my good fortune that she, so infinitely my superior in every single moral quality, consented to be my wife." (Darwin Museum, Down House)

Anne Elizabeth Darwin, Charles's eldest daughter, who died in 1851 at age ten. Daguerreotype taken in 1849. (Darwin Archives, Cambridge University)

The birth of another son, Horace (1851–1928), on 13 May 1851, lifted the family's spirits, and slowly Darwin resumed his barnacle work. Horace later founded the Cambridge Instrument Co., was elected a Fellow of the Royal Society, and was knighted in 1918. In 1853, Charles was awarded the Royal Medal for volume 1 of his barnacle monograph, and barnacles occupied his time completely until 1854.

This zoological work resulted in contact with Thomas Henry Huxley, who became Darwin's friend, confidant, and ideological champion. Secular, professional scientists such as Huxley, utilizing the naturalism of the scientific method, were beginning to challenge the creationist, Anglican establishment that still relied on "revelation" to explain nature. This developing enlightenment allowed Darwin's ideas to flourish on their own merit.

After eight years of barnacles, Darwin was eager to move on to another field of study that would provide even more evidence for his theory of evolution—the domestication of animals and plants and the breeding of fancy pigeons, the latter being a popular hobby at the time. He built a pigeon coop in the garden and began to breed all sorts of forms. In addition, he boiled the bones of bird corpses acquired from other pigeon-fanciers, in order to study osteological variations. He soon came to realize that if man could create such divergent forms of pigeons from the common rock dove in only several generations, what could nature do given thousands of years? He further observed that in all these highly selected varieties of pigeons, the embryos of all strains more closely resembled the ancestral rock dove than the adults did.

Darwin also began a series of long-running experiments on the dispersal and viability of seeds that he soaked in sea water. Even after forty-two days of immersion, some seeds

Darwin family outside a bay window at Down House, about 1863. *Left to right*: Leonard, Henrietta, Horace, Emma, Elizabeth, Francis, and a school friend called Spitta. (Cambridge University Library)

Thomas Henry Huxley, "Darwin's Bulldog." He served as a surgeon on the HMS *Rattlesnake*, mostly in Australian waters, from 1846 to 1850. He was made a Fellow of the Royal Society in 1850 and was a frequent correspondent with Charles, as well as a frequent guest at Down House. (Portrait from a photograph by Elliott and Fry. Steel engraving in *Nature*, 5 February 1874.)

could still germinate, a period that was long enough to allow plants to colonize oceanic islands by drifting on currents. He published a series of papers in the *Gardener's Chronicle* about this and many other botanical topics.

Financially, the Darwins were well off. Early on, with Dr. Darwin's advice, Charles and Emma had invested in railway and canal stock. Charles even thought about emigrating to Australia to secure his children's financial future. Speculation was something he learned from his wealthy father, whose estate totaled £223,759. Charles inherited £51,000 at Robert Darwin's death, and this enabled him to pursue his pure research on barnacles without any need to earn a living for the rest of his life.

ELEVEN

 *Alfred Russel Wallace and* The Origin

IN 1855, LYELL SUGGESTED that Darwin read a paper by an unknown naturalist, entitled "On the law which has regulated the introduction of new species." The author was Alfred Russel Wallace. It showed some similarities to Darwin's own sketches about evolution and natural selection in 1842 and 1844, but Darwin did not recognize Wallace's writings as a threat. Lyell thought otherwise and urged Charles to prepare a summary of his theory. So Charles set to work on his "big book," which he was calling "Natural Selection." Portions of the manuscript were read by Hooker, Huxley, and John Lubbock, Darwin's friend, neighbor, and a member of Parliament. On 5 September 1857, Darwin sent a summary of the "big book" to Asa Gray, his American botanist friend at Harvard.

In 1858, while recovering from malaria in the Malay Archipelago, Wallace read, as Darwin did eighteen years earlier, Malthus' *An Essay on the Principle of Population.* Wallace reached the same conclusion as Darwin, namely, that natural selection was the factor controlling populations. Wallace wrote an essay, "On the Tendency of Varieties to Depart Indefinitely from the Original Type," and sent it to Darwin. He asked Darwin to arrange for its publication if the ideas were sound.

When he received Wallace's essay on 18 June 1858, Darwin

Alfred Russel Wallace, age twenty-five in 1848 (*left*) and age fifty-five in 1878 (*right*). He was a traveler and naturalist and a co-discover, with Darwin, of natural selection. Wallace collected natural history specimens in the Amazon (1848–52) and in the Malay Archipelago (1854–62). He is best known for establishing biogeographical realms and for "Wallace's Line," separating the Asian realm (west) from the Australian realm (east). He was a frequent visitor to Down House. He was the first recipient of the Darwin Medal in 1890 and accepted appointment as a Fellow of the Royal Society in 1905. (From Wallace [1905])

was stunned. He had already written a quarter of a million words of his "big book," which was expected to fill three volumes. Wallace's ideas were identical to what he had been working on for twenty years. Darwin wrote to Lyell, saying "I never saw a more striking coincidence; if Wallace had my Ms sketch written out in 1842, he could not have made a better abstract!"

The timing was very bad for Darwin, as two of his children were seriously ill. Henrietta was recovering from diphtheria, and Charles Waring (1856–1858), the youngest child,

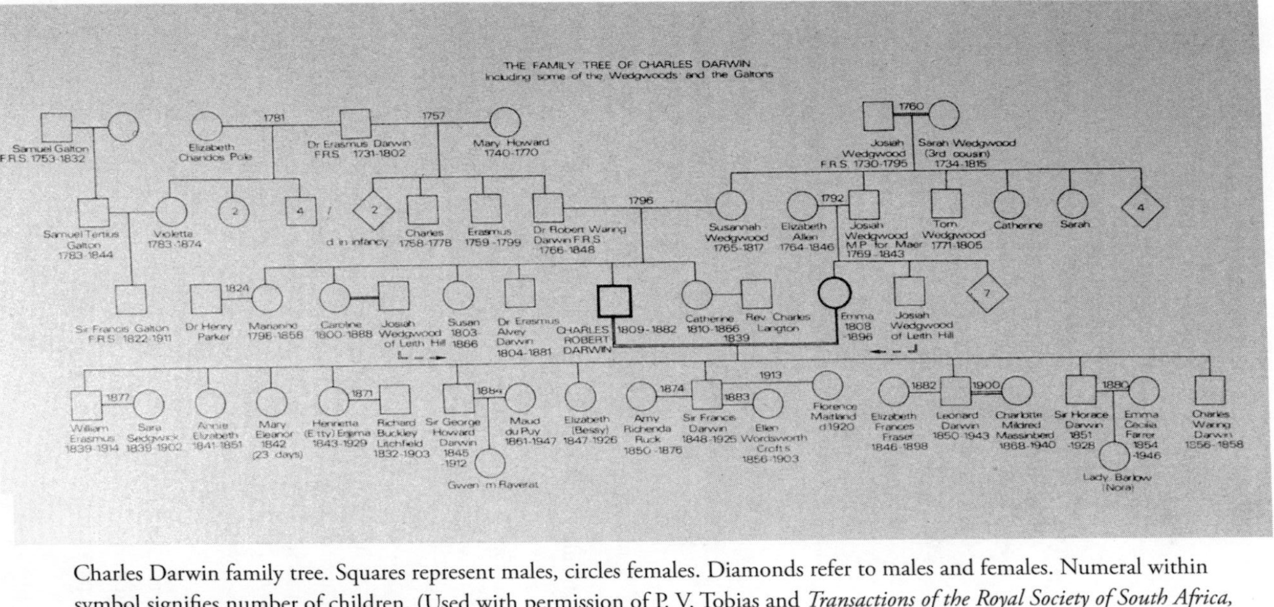

Charles Darwin family tree. Squares represent males, circles females. Diamonds refer to males and females. Numeral within symbol signifies number of children. (Used with permission of P. V. Tobias and *Transactions of the Royal Society of South Africa,* the journal in which his 1972 article appeared.)

Charles Lyell (*standing*) and Joseph Dalton Hooker (*right*) advise Darwin regarding Wallace's letter. (From a scene at Down House recreated by a Russian artist, variously referred to as Victor Eustaphieff or Evstafieff, for the centenary of *The Origin*.)

was mortally ill with scarlet fever and died a few days later. Emma was forty-eight when Charles Waring was born. He never learned to walk or talk, and he may have been a victim of Down syndrome.

Darwin was desperate to decide whether or not it was ethical to publish before Wallace. The idea that his life's work would be in vain was devastating, but he was so consumed with grief for his children that he was nearly paralyzed. He wrote to Hooker on 29 June: "I am quite prostrated and can do nothing but I send Wallace and my abstract of my letter to Asa Gray . . . [and] my sketch of 1844 . . . I really cannot bear to look at it. Do not waste much time. It is miserable in me to care at all about priority."

Lyell and Hooker were, of course, privy to Darwin's work for the past twenty years on the idea of evolution by means of natural selection. They were determined to prevent Darwin from being scooped and persuaded him to prepare a paper that would be presented, along with Wallace's paper, at a meeting of the Linnean Society. These two papers—"On the Tendency of Species to Form Varieties" and "On the Perpetuation of Varieties and Species by Means of Selection"—were read to a group of about thirty people on 1 July 1858 and were subsequently published in the *Journal of the Proceedings of the Linnean Society (Zoology)*.

When Wallace finally learned of this event, he was very generous and recognized that all the real work had been done by Darwin. Later, in his 1905 autobiography, Wallace stated:

> Both Darwin and Dr. Hooker wrote to me in the most kind and courteous manner, informing me of what had been done, of which they hoped I would approve. Of course I not only approved, but felt that they had given me more honour and credit than I deserved, by putting my sudden intuition . . . on the same level with the prolonged labours of Darwin, who had reached the same point twenty years before me, and had worked continuously during that long period in order that he might be able to present the theory to the world with such a body of systematized facts and arguments as would almost compel conviction. (*My Life*, 193)

Darwin and Wallace remained in contact throughout Darwin's life, and Charles even arranged for a government pension for Wallace. At the time, few people recognized the significance of the joint papers. Darwin abandoned his "big book" and began to prepare what he called an "abstract" of his species theory on 20 July 1858. This took him months of working for several hours each day while continuing to deal

[*From the* JOURNAL *of the* PROCEEDINGS OF THE LINNEAN SOCIETY *for August* 1858.]

On the Tendency of Species to form Varieties; and on the Perpetuation of Varieties and Species by Natural Means of Selection. By CHARLES DARWIN, Esq., F.R.S., F.L.S., & F.G.S., and ALFRED WALLACE, Esq. Communicated by Sir CHARLES LYELL, F.R.S., F.L.S., and J. D. HOOKER, Esq., M.D., V.P.R.S., F.L.S., &c.

[Read July 1st, 1858.]

London, June 30th, 1858.

MY DEAR SIR,—The accompanying papers, which we have the honour of communicating to the Linnean Society, and which all relate to the same subject, viz. the Laws which affect the Production of Varieties, Races, and Species, contain the results of the investigations of two indefatigable naturalists, Mr. Charles Darwin and Mr. Alfred Wallace.

These gentlemen having, independently and unknown to one another, conceived the same very ingenious theory to account for the appearance and perpetuation of varieties and of specific forms on our planet, may both fairly claim the merit of being original thinkers in this important line of inquiry; but neither of them having published his views, though Mr. Darwin has for many years past been repeatedly urged by us to do so, and both authors having now unreservedly placed their papers in our hands, we think it would best promote the interests of science that a selection from them should be laid before the Linnean Society.

Taken in the order of their dates, they consist of:—

1. Extracts from a MS. work on Species*, by Mr. Darwin, which was sketched in 1839, and copied in 1844, when the copy was read by Dr. Hooker, and its contents afterwards communicated to Sir Charles Lyell. The first Part is devoted to "The Variation of Organic Beings under Domestication and in their Natural State;" and the second chapter of that Part, from which we propose to read to the Society the extracts referred to, is headed, "On the Variation of Organic Beings in a state of Nature; on the Natural Means of Selection; on the Comparison of Domestic Races and true Species."

2. An abstract of a private letter addressed to Professor Asa Gray, of Boston, U.S., in October 1857, by Mr. Darwin, in which

* This MS. work was never intended for publication, and therefore was not written with care.—C. D. 1858.

Title page of Darwin's and Wallace's papers, read before the Linnean Society on 1 July 1858 and published in the *Journal of the Proceedings* in August. This solution, organized by Lyell and Hooker, preserved Darwin's priority but also allowed Wallace to receive credit for his independent development of the concept of natural selection. Darwin's parts included a portion of his 1844 essay and part of his summary letter to Asa Gray in 1857.

ON

THE ORIGIN OF SPECIES

BY MEANS OF NATURAL SELECTION,

OR THE

PRESERVATION OF FAVOURED RACES IN THE STRUGGLE FOR LIFE.

BY CHARLES DARWIN, M.A.,

FELLOW OF THE ROYAL, GEOLOGICAL, LINNÆAN, ETC., SOCIETIES;
AUTHOR OF 'JOURNAL OF RESEARCHES DURING H. M. S. BEAGLE'S VOYAGE ROUND THE WORLD.'

LONDON:
JOHN MURRAY, ALBEMARLE STREET.
1859.

The right of Translation is reserved.

Title page from the first edition of *The Origin of Species*, published on 24 November 1859. This date marks the beginning of modern biology.

with his constant gastric distress. This 155,000-word "abstract" was *The Origin of Species.*

The manuscript was finished by mid-March 1859. Lyell suggested that Darwin contact John Murray to publish the book, and Murray agreed without even seeing the manuscript. Murray and Darwin became good friends, and Murray eventually published first editions of ten of Darwin's books. *On the Origin of Species by Means of Natural Selection, or the Preservation of Favoured Races in the Struggle for Life* was published on 24 November 1859. This date marks the beginning of modern biology (which will celebrate its 150th anniversary in 2009).

Murray received 1500 orders when the book was released, two days before its official date of publication. This is why you often hear it said that *The Origin* sold out on its initial day of publication. Murray immediately asked Darwin to prepare a second edition, and 3000 copies were sold soon afterwards. The book went through six editions and, like Darwin's *Journal of Researches* [*Voyage of the Beagle*], has never been out of print.

Wallace commented on *The Origin* in a letter to Henry Walter Bates, dated 24 December 1860:

> I know not how, or to whom, to express fully my admiration of Darwin's book. To *him* it would seem flattery, to others self-praise; but I do honestly believe that with however much patience I had worked and experimented on the subject, I could *never have approached* the completeness of his book, its vast accumulation of evidence, its overwhelming argument, and its admirable tone and spirit. I really feel thankful that it has not been left to me to give the theory to the world. Mr. Darwin has created a new science and a new philosophy; and I believe that never has such a complete illustration of a new branch of human knowledge been due to the labours and researches of a single man.

Darwin as an ape, drawn in response to publication of *The Descent of Man*. (From *Hornet*, 22 March 1871)

Richard Owen, on the other hand, anonymously wrote a scathing review of *The Origin* in April 1860, more out of spite and envy than biology. This ended his and Darwin's friendship. *The Origin* energized not only scientists but the general public as well. Numerous newspapers and magazines caricatured Darwin as an ape, even though he avoided direct discussion of this issue in *The Origin*. However, he did put humans on a level with other animals and concluded by writing that pregnant sentence, "Much light will be thrown on the origin of Man and his history."

TWELVE

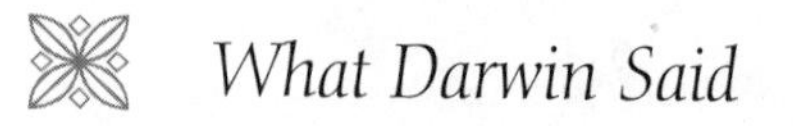

What Darwin Said

FROM HIS BREEDING EXPERIMENTS and observations in nature, Darwin recognized that many more offspring were produced than actually survived. This applied to plants and animals—trees to elephants, and everything in between. Some animals produce millions of eggs or thousands of larvae, and some plants produce millions of spores or seeds, but, fortunately, the vast majority of this overproduction does not survive to adulthood.

Darwin's patience and keen powers of observation led to the realization that there is variation in nature. No two individuals are alike in a litter of puppies, school of minnow hatchlings, or members of the same species of barnacles or orchids. The germination of seeds from the same plant yields variable offspring. Darwin's genius was to understand that this overproduction was related to variation. He eventually came to the realization that there is competition for resources in nature and that the variations best adapted to their environment would displace the less-favorably endowed individuals.

Since the environment is doing the choosing, he called this process *natural selection*, as opposed to the artificial selection imposed by breeders. This resulted in *descent with modification*, which was his definition (and still a perfectly good one) of *evolution*. Today we have the benefit of genetic knowledge, which was unknown to Darwin, in comprehending how this process works. Descent with modification can be explained as a *change in gene frequency*, that is, a change in the proportion of a particular gene variant among all the alternative forms of that gene. Natural selection is *differential reproduction.* In other words, in the same environment, one form leaves more offspring than another form. The *environment* is the selecting agent.

Darwin had no knowledge of the source of this variation, and there was no way he could know that the "sports" he observed in breeding experiments were chemical mistakes, a sudden inheritable change in a gene (*mutations*). Today we understand that genetic variation is produced by mutation, sexual reproduction, chromosome rearrangement, etc.

So, to recap: *evolution is descent with modification (change in gene frequency), brought about by natural selection (differential reproduction), acting on the variations produced by mutation and other sources, with the environment doing the selecting.* How elegantly simple! As Thomas Henry Huxley wrote upon receiving a copy of *The Origin*, "How extremely stupid of me not to have thought of that."

THIRTEEN

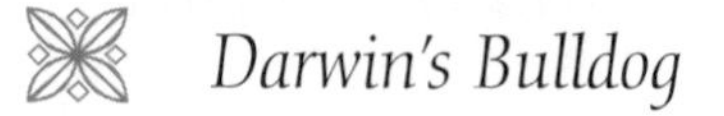

Darwin's Bulldog

The Origin BECAME THE MAIN TOPIC of conversation in scientific circles, but Darwin remained secluded at Down. The first major test of how the scientific community would view his theory was at the British Association for the Advancement of Science meeting in Oxford on 30 June 1860. Darwin did not attend, knowing that his supporters would champion his cause. The Bishop of Oxford, Samuel Wilberforce, known as "Soapy Sam" because of his slick oratory, was invited to reply to a paper on "civilization according to the Darwinian hypothesis." It was obvious that Wilberforce had been coached by Richard Owen.

The meeting was crowded with about 1000 people, including reporters and politicians, and tension filled the air. The bishop paused during his monologue, turned to Huxley (who was representing Darwin), and asked whether it was on his grandfather's or his grandmother's side that he was descended from an ape. The audience erupted with laughter at this insult. Huxley slowly arose, turned to the bishop, and replied to the effect that given the choice of a lowly ape or a person of privilege such as my lord bishop who would introduce ridicule into a scientific discussion, he would unhesitatingly choose the ape! Pandemonium broke out, women swooned, undergraduates chanted "monkey, monkey," and FitzRoy stalked out of the room holding a bible over his head

The two combatants in the Oxford debate of 1860: Reverend Samuel Wilberforce, Bishop of Oxford (*left*), and Dr. Thomas Henry Huxley (*right*). (From cartoons in *Vanity Fair*, July 1869 and January 1871, respectively.)

and shouting "The Book! The Book!" This must have been an incredible moment and would make a wonderful movie scene. This is how Huxley earned the nickname "Darwin's Bulldog."

The Origin was translated into eleven European languages within Darwin's lifetime and at least twenty-nine languages to date. The first Japanese edition was published in 1896, and the first Chinese edition in 1903.

FOURTEEN

 A Man of Enlarged Curiosity

DARWIN DID NOT QUIETLY RETIRE after publication of *The Origin,* although he suffered greatly as his health continued to decline. Instead, he turned his attention to botany. By studying the morphology and anatomy of flowers, especially orchids, Darwin discovered that nearly all parts of a flower are adapted to allow pollination by insects (as opposed to the creationist belief that flowers were designed to provide beauty for humans).

He published *On the Various Contrivances by which Orchids are Fertilized by Insects* in 1862. In this book, he expressed amazement at an orchid (*Angraecum sesquipedale*) from Madagascar that has an 11½ inch long, whiplike nectary (a gland that secretes nectar) and speculated that there must be a moth in that island with a proboscis (a flexible snoutlike body part used to extract nectar) capable of reaching something this deep. It was not until 1903 that entomologists described a giant hawk-moth from Madagascar with such an astonishing proboscis; they named it *Xanthopan morganii praedicta* in honor of Darwin's prediction.

Darwin was greatly influenced by Asa Gray, his American botanist friend and correspondent at Harvard University, who had written a paper on the coiling tendrils of plants. Darwin conducted experiments on this subject in his greenhouse at Down. *The Movements and Habits of Climbing Plants* was published by the Linnean Society in 1865 and reissued as a book

Greenhouse at Down where Darwin conducted his botanical experiments. (Author's photograph, 1974)

in 1875 by his publisher, John Murray. Darwin's experiments were often ingenious, and carried out meticulously to control all factors. In fact, because of his botanical experiments, the word "control" entered the literature of experimental biology.

Next, he gathered together his vast database of facts and observations on domestication and breeding, especially his own work with pigeons. Some data were already organized as chapters in his unpublished "Natural Selection" big book, and this material was published in 1868 in his two-volume *The Variation of Animals and Plants under Domestication*. Herbert Spencer's phrase "survival of the fittest" was first used by Darwin in *Variation*, and he repeated it in the fifth edition of *The Origin* in 1869. *Variation* also elaborated on Darwin's view of heredity. He suggested, erroneously, that particles from the body accumulate in the ovaries and testes and that these "gemmules" are transmitted to form offspring and account for inherited characteristics.

The Descent of Man and Selection in Relation to Sex was published in 1871. In *The Descent*, Darwin used the term "evolution" for the first time in its modern sense. Later, in the sixth and final edition of *The Origin* in 1872, he more fully discussed his "transmutation" theory in terms of "evolution." *The Descent* is essentially two works. The first third is a continuation of *The Origin*, as applied to man. There were virtually no human fossils known at this time, except for a few confusing Neanderthal bones for which there had not yet been a good explanation. No australopithecines (ape-men, or "missing links") were known. Yet Darwin correctly reasoned, on biogeographical grounds, that Africa was the cradle of human evolution. This has since been confirmed by an abundance of *Australopithecus* species found only in Africa and by DNA data.

The brilliance of his reasoning is shown in the following quote from *The Descent of Man*: "In each great region of the world the living mammals are closely related to the extinct species of the same region. It is therefore probable that Africa was formerly inhabited by extinct apes closely allied to the gorilla and chimpanzee; and as these two species are now man's nearest allies, it is somewhat more probable that our early progenitors lived on the African continent than elsewhere."

The second two-thirds of the book, sexual selection, explained the development of secondary sexual characters in one sex by the preferences of the other sex—for example, peahens choosing to mate with peacocks with elaborate tails. *The Descent* re-ignited the volatile debate begun by *The Origin*, because it clearly stated what *The Origin* had only hinted at: that humankind was directly descended from animals. Darwin himself was relatively unaffected by the public controversy, remaining secluded at Down and continuing to gather evidence to support his theory.

After finishing *The Descent*, Darwin began drawing together all his observations on emotion, including the notes he made when his children were babies, about crying, smiling, etc. *The*

Cartoon of Darwin as a monkey, just after *The Expression of the Emotions in Man and Animals* was published. Many other similar personal attacks were published during his lifetime. (From *Fun*, 16 November 1872)

Expression of Emotions in Man and Animals, a pioneering study of animal behavior, was published in 1872. It is also one of the first books in which photographs were used to illustrate the text. This particular work earned more money for Darwin than any of his other books. It also demonstrated that Darwin erroneously accepted the idea that acquired characteristics were inherited, which was the prevailing view of most scientists of the day.

There seems to have been no end to Darwin's creative

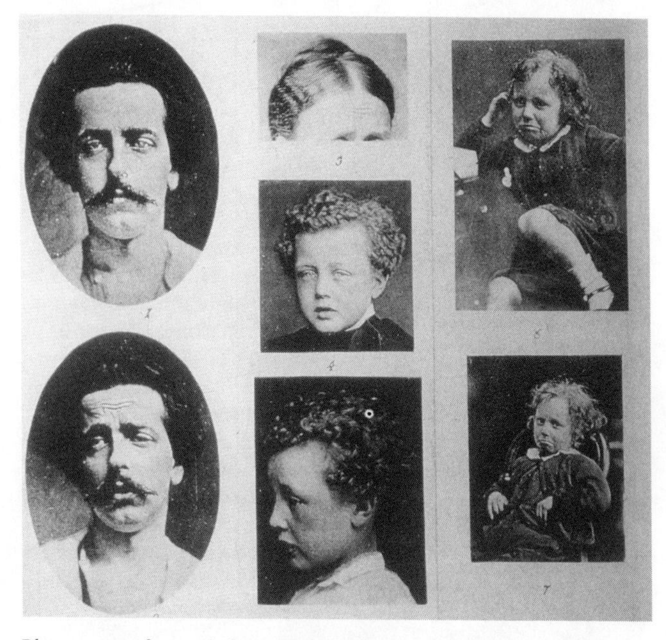

Plate 2, one of seven heliotypes from *Expression of the Emotions.* This was one of the first books to use photographs.

genius. During the last ten years of his life, Darwin continued to circle the Sandwalk every day and spend many hours observing plants and animals in his garden. Darwin's work habits and celebrity notwithstanding, the Down House gardener, Henry Lettington, observed him staring at a flower for ten minutes and remarked, "If only he had something to do I believe he would be better."

Darwin's various botanical books were widely praised by professional botanists, but, because they were highly technical, not all of them were popular with the public. *Insectivorous Plants,* published in 1875, focused on his experiments with the insect-eating *Drosera* (or sundew) that he had begun fifteen years earlier. It explained how insectivorous plants survive in nitrogen-deficient soils by digesting insects, whose bodies are high in nitrogen.

The Effects of Cross- and Self-Fertilisation in the Vegetable Kingdom was published in 1876 and complemented his orchid book. By counting and weighing seeds produced by various crosses, Darwin showed that the offspring of cross-fertilized individuals were more vigorous than the offspring of self-fertilized plants. All the experiments were conducted in the specially designed beds and greenhouses at Down by Darwin, his staff, and his children. The results of this work caused Darwin to worry about the long-term health of his children. He and Emma were first cousins, so he feared that their offspring might inherit weaknesses (from what could be considered a kind of self-fertilization) that would not be present

Darwin on his horse Tommy, in the late 1870s. Darwin rode a great deal, both as a young man and when in South America. He took it up again later in life as a form of exercise, until his horse stumbled and Darwin fell. The long exposure for this photograph is evident from Tommy's head movement. (Darwin Archives, Cambridge University Library)

One of Darwin's calling cards. (William L. Clements Library, University of Michigan)

in the children of two unrelated (and hence cross-fertilizing) individuals.

Darwin's son Francis (Frank) moved back to Down after his first wife's death in 1876, and worked with his father on research projects. *The Different Forms of Flowers on Plants of the Same Species* was published in 1877 and dedicated to Asa Gray. In this book, Darwin studied primroses and confirmed that self-fertilization was not as effective as cross-fertilization.

In *The Power of Movement in Plants*, published with Francis in 1880, Darwin discovered that the growing tip of a shoot is light sensitive and that growth is greatest on the side of the stem away from the light. These discoveries opened up the field of plant growth hormones.

The Formation of Vegetable Mould Through the Action of

Worms With Observations on Their Habits, published in 1881 with much help from Frank, was the culmination of Darwin's longest-running sequence of observations. This was his last book and the most popular of any of Darwin's titles. Darwin's son Horace, an engineer, designed an instrument to measure the vertical movements of a specially positioned wormstone. In a classic study of quantitative ecology, they calculated that earthworms bring eighteen tons of finely ground soil per acre per year to the surface, thereby aerating and improving the soil.

Darwin kept worms in pots in his study and watched them at night with light from a lantern covered with red glass. He also enlisted the entire family to test if worms

Darwin standing on the veranda at Down House in 1881, at age seventy-two. (Photograph by Elliott and Fry, London. Darwin Archive, Cambridge University Library)

Emma Darwin in 1881, at age seventy-three. (Darwin Museum, Down House)

could hear. Frank's son Bernard played a metal whistle, Frank the bassoon, and Emma the piano. The worms proved to be deaf, but they did react to the vibrations of the piano when they were placed on it. One of the last acts of Darwin's life was to make revisions to the sixth printing of the worm book, which was published in early 1882. This book is still considered the starting point among commercial worm growers.

Darwin had received the Copley Medal of the Royal Society in 1864, and in 1877, his old school, Cambridge University, awarded him an honorary doctorate. In his last years, Darwin's health improved somewhat, and he received many honors and much recognition. Nevertheless, he was far too controversial to be knighted by Queen Victoria.

FIFTEEN

Darwin's Death

FRANCIS DARWIN SUMMARIZED his father's religious views in *Life and Letters* (1897), and Atkins (1974) provided additional insight. Charles considered such things to be very personal and took great pains to avoid offending Emma and her beliefs. Darwin began life in a freethinking, nominally Anglican-Unitarian environment and was never seriously religious. He studied theology during his days at Cambridge, but the more he learned about how nature operated, the less he accepted the revealed religion of the Bible. Ever the scientist, he gave up Christianity because "it is not supported by evidence." Nor could he accept the notion that unbelievers such as his father, his brother, and most of his close friends would be condemned to suffer forever in hell.

The loss of his father and, more importantly, his daughter Annie's death at age ten, reinforced his biological observations that all life is involved in a struggle against disease, famine, predation, and death, and that there is no providential force acting on the human condition any more than there is one acting on other parts of nature. He never specifically denied the existence of a God, but if he accepted such a concept, it was as a distant, impersonal force that operated through material cause and effect—a natural, lawgiver sort of God. The word "agnostic" was coined by his friend Thomas Henry Huxley in 1864, and Darwin himself wrote that "agnostic would be the more correct description of my state of mind."

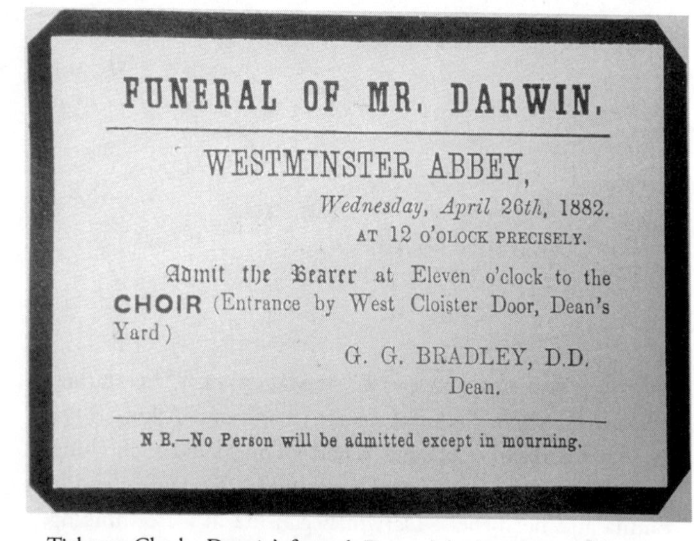
FUNERAL OF MR. DARWIN.

WESTMINSTER ABBEY,

Wednesday, April 26th, 1882.

AT 12 O'CLOCK PRECISELY.

Admit the Bearer at Eleven o'clock to the **CHOIR** (Entrance by West Cloister Door, Dean's Yard)

G. G. BRADLEY, D.D.
Dean.

N.B.—No Person will be admitted except in mourning.

Ticket to Charles Darwin's funeral. Darwin's body was carried from Down House in a hearse drawn by four black horses. (Darwin Archives, Cambridge University Library)

Although Darwin's *Autobiography* (ed. Barlow 1958) is essential reading for any Darwinophile, his recollections were written for his children, without any thought that they would ever be published. These sketches were written in 1876, added to in 1878, and updated in 1881.

The last months and days of Darwin's life are chronicled by Atkins (1974). On Christmas Day in 1881, he suffered chest pains, which increased as he shuffled around the Sandwalk in February and March of 1882. Additional attacks occurred on the 4th and 5th of April. Various doctors came and went, including Dr. Andrew Clark, physician to Queen Victoria. Clark refused to accept a fee and considered it a privilege to attend to such a famous person. During the night of 18 April 1882, Darwin suffered a severe heart attack and lost consciousness; he was revived with difficulty. He seemed to recognize the approach of death, and he said, "I am not the least afraid to die." All the next morning he suffered from terrible nausea and faintness. He died at about 4:00 p.m. on 19 April 1882, in the seventy-third year of his life. He was attended by Emma and his children Francis, Henrietta, and Elizabeth.

Emma wanted a simple funeral in the ancient graveyard adjoining the little church at Downe, but at the request of twenty members of Parliament headed by John Lubbock, Charles Darwin was laid to rest with pomp and ceremony in Westminster Abbey on 26 April 1882. Pallbearers included Hooker, Huxley, Wallace, Lubbock, the president of the Royal Society, the American ambassador, the chancellor of Cambridge University, and other notable men. The funeral was attended by all surviving Darwin children and the family servants, as well as representatives from the United States, France, Germany, Italy, Spain, and Russia; others from universities and every scientific society in Great Britain; and a large number of personal friends and distinguished men. Emma, however, did not attend. His grave is in the northeastern corner of the nave, a few feet from that of Isaac Newton and Charles Lyell.

Darwin's funeral at Westminster Abbey. Darwin was the first and only naturalist to be buried in Westminster Abbey. He was buried in a coffin of white oak. (From the *Graphic*, 6 May 1882, p. 1)

One of the last photographs of Charles Darwin, by Elliott and Fry. (Down House)

Epilogue

OCCASIONALLY A STUDENT from a fundamentalist background will come up to me after an evolution class and say, "Did you know that Darwin recanted his theory before he died?" Or an ad for a creationist lecture will say something about a "deathbed conversion" of Darwin. These apocryphal stories represent wishful thinking on the part of the anti-evolutionists, but they do have an interesting history. This myth can be traced to a British evangelist called Lady Hope, who addressed a religious meeting in Boston in 1915 or 1916. She claimed to have visited Darwin near the time of his death and found him reading the bible and singing hymns.

These stories proliferated in newspapers around the world, to the extent that Darwin's daughter Henrietta, then Mrs. Litchfield, issued a statement to the *Christian* that the paper published "in the interest of truth" on 23 February 1922. Henrietta, who was present at Darwin's death, stated that "he never recanted any of his scientific views . . . and the whole story has no foundation whatever." Francis, who was also at his father's bedside, denied that a Lady Hope ever visited Down. See Atkins (1974), the two papers by Sloan (1960, 1965), and the book by James Moore (1994) for an exposition of this fraud.

The publication of Darwin's *On the Origin of Species* in 1859 created a paradigm shift from creation to evolution. Darwin showed that humans are part of nature, not above it, and that all animal life, including human, is related by descent from a common ancestor. His mechanism of evolution via natural selection is a powerful creative force that provided an explanation for the diversity of life. This dramatic change in world view from supernaturalism to methodological naturalism has allowed staggering scientific advances in the past 150 years which transcend science and permanently impact on the human psyche.

For me personally, Darwin's name evokes a stream of wonderful memories, such as being the sole visitor to Down House one lovely summer's day or climbing over volcanic rocks as I led a tour through the Galápagos Islands. Some of the happiest days of my life were spent in his namesake city, Darwin, Australia—that most beautiful, exotic, diverse, and friendly of tropical cities where I did nurseryfish field research on the crocodile-rich Adelaide River. The City of Darwin is also the home of Charles Darwin University, a fitting tribute to one of the most influential scientists who ever lived.

The late Theodosius Dobzhansky, eminent geneticist and major contributor to the modern synthesis of genetics and evolution, famously wrote: "Nothing in biology makes sense except in the light of evolution." That light was first lit in England, some two centuries ago, by an extraordinary man we now refer to simply as "Darwin."

APPENDIX A

Books

DARWIN SCHOLARS are indebted to R. B. Freeman for his annotated bibliographic list, *The Works of Charles Darwin* (1977).

The Zoology of the Voyage of H.M.S. Beagle, *under the Command of Captain Fitzroy, During the Years 1832 to 1836.* Edited and superintended by Charles Darwin. Smith Elder, London. 5 parts. 1838–43. 632 pp., including 166 plates.

Darwin's authors were Richard Owen (Part 1—Fossil *Mammalia*), George Robert Waterhouse (Part 2—*Mammalia*), John Gould (Part 3—Birds), Leonard Jenyns (Part 4—Fish), and Thomas Bell (Part 5—Reptiles). Darwin superintended the production of this work, contributed introductions to parts 1 and 2, added notes of habits and ranges to parts 2 and 3, and allowed his collection labels to be used in parts 4 and 5.

Narrative of the Surveying Voyages of His Majesty's Ships Adventure *and* Beagle, *etc.* Edited by Robert FitzRoy. Henry Colburn, London. 3 vols. 1839.

Volume 1 was authored by Captain King, and volume 2 was written by Captain FitzRoy. Volume 3 was written by Charles Darwin and was entitled *Journal and Remarks, 1832–1836.* It was 615 pages long and is what is now universally known as *The Voyage of the Beagle*. In 1839, Henry Colburn published this book in its own right as *Journal of Researches into the Geology and*

Natural History of Various Countries Visited by H.M.S. Beagle *Round the World Under the Command of Capt. Fitzroy, R. N.* In 1845, John Murray published a second edition and modified the title to *Journal of Researches into the Natural History and Geology and of Various Countries Visited by H.M.S.* Beagle *Round the World Under the Command of Capt. Fitzroy, R. N.* An 1860 reprinting of the second edition by John Murray is considered the definitive edition of this work. The title *The Voyage of the Beagle* was first used in a 1905 printing by Harmsworth Library, London.

The Structure and Distribution of Coral Reefs: Being the First Part of the Geology of the Voyage of the Beagle, *under the Command of Capt. Fitzroy, R. N. During the Years 1832–1836.* Charles Darwin. Smith Elder, London. 1842. 214 pp. 2nd ed. 1874; 3rd ed. 1889.

Darwin reasoned that coral reefs and atolls resulted from upward coral growth that matched the subsidence of the ancient volcanoes upon which the corals grow. This theory was confirmed in the 1950s, when drilling on Pacific atolls revealed volcanic rock hundreds of meters below the surface.

Geological Observations on the Volcanic Islands Visited During the Voyage of H. M. S. Beagle, *Together with Some Brief Notices of the Geology of Australia and the Cape of Good Hope: Being the Second Part of the Geology of the Voyage of the* Beagle, *under the Command of Capt. Fitzroy, R. N. During the Years 1832–1836.* Charles Darwin. Smith Elder, London. 1844. 175 pp.

Geological Observations on South America: Being the Third Part of the Geology of the Voyage of the Beagle, *under the Command of Capt. Fitzroy, R. N. During the Years 1832–1836.* Charles Darwin. Smith Elder, London. 1846. 279 pp.

Parts 2 and 3 of *The Geology of the Beagle* discussed earth movements and the minerals in granites and lava. They also are where Darwin originated the deformation theory of metamorphic rock.

Geological Observations on Coral Reefs, Volcanic Islands, and on South America. Charles Darwin. Smith Elder, London. 1851.

In this work, the three volumes of *The Geology of the Beagle* (published in 1842, 1844, and 1846) were incorporated into one volume, as originally intended. In 1876, Smith Elder published a second edition, containing only *Volcanic Islands* and *South America.* A third edition appeared in 1891.

A Monograph of the Sub-Class Cirripedia, with Figures of All Species. Volume 1: *The Lepadidae; or Pedunculated Cirripedes.* Charles Darwin. Ray Society, London. 1851. 400 pp. Volume 2: *The Balanidae (or Sessile Cirripedes), etc.* Charles Darwin. Ray Society, London. 1854. 684 pp. *A Monograph of the Fossil Lepadidae or Pedunculated Cirripedes of Great Britain,* vol. 1. Charles Darwin. Paleontographical Society, London. 1851. 88 pp. *A Monograph of the Fossil Balanidae and Verrucidae*, vol. 2. Charles Darwin. Paleontographical Society, London. 1854. 44 pp. Index [to all four volumes], 1858.

These four volumes on recent and fossil barnacles—still highly esteemed by barnacle taxonomists—cost Darwin eight years (1846–53) of tedious dissection and examination. They are his main excursion into taxonomy and undoubtedly honed his keen eye for detecting variations.

On the Origin of Species by Means of Natural Selection, or the Preservation of Favoured Races in the Struggle for Life. Charles Darwin. John Murray, London. 1859. 502 pp. 2nd ed. 1860; 3rd ed. 1861 (538 pp.); 4th ed. 1866 (593 pp.); 5th ed. 1869 (596 pp.); 6th ed. 1872 (458 pp.).

The last edition, set in smaller type, was extensively revised and included a new chapter. Darwin used the word "evolution" in this edition; previous editions used "evolved." ("Evolution" was first used in *The Descent of Man* in 1871.) The 1876 printing of the sixth edition is considered the first issue of the definitive text.

On the Various Contrivances by which British and Foreign Orchids Are Fertilised by Insects and on the Good Effects of Intercrossing. Charles Darwin. John Murray, London. 1862. 365 pp. Title simplified to *The Various Contrivances by which Orchids Are Fertilised by Insects* for 2nd ed. 1877. 300 pp.

Darwin correctly explained that flower color and shape created an attractive landing pad for pollinating insects.

On the Movements and Habits of Climbing Plants. Charles Darwin. *Journal of the Proceedings of the Linnean Society of London* 9 (33–34): 1–128. 1865.

The second edition of this lengthy journal article (usually considered the first edition in book form) was published by John Murray in 1875. Darwin explained that the twining movements of plants were an adaptation whereby the plant would be exposed to more sunlight. The mechanism for this process was described in *The Power of Movement in Plants.*

The Variation of Animals and Plants under Domestication. Charles Darwin. John Murray, London. 1868. 2 vols.: vol. 1, 411 pp. and vol. 2, 486 pp. 2nd ed. 1875. Extensively revised for the second edition.

Darwin used economist Herbert Spencer's expression "survival of the fittest" for the first time. He recorded a bewildering amount of variation found in domestic plants and animals and elaborated his erroneous and complicated hypothesis of pangenesis. This work may be considered a complete account of the material Darwin abstracted for the first chapter of *The Origin.*

The Descent of Man and Selection in Relation to Sex. Charles Darwin. John Murray, London. 2 vols.: vol. 1, 1870 and vol. 2, 1871. 2nd ed. (vols. 1 and 2 combined) 1874.

This is essentially two works. The shorter *Descent of Man*—the first of Darwin's books to use the word "evolution"—is the full exposition of Darwin's statement in *The Origin* that "light would be thrown on the origin of Man and his history." *Selection in Relation to Sex* explains that sexual selection, a special form of natural selection, is a result of variations that affect the ability to attract mates.

The Expression of Emotions in Man and Animals. Charles Darwin. John Murray, London. 1872. 374 pp. 2nd ed. 1890 (394 pp.). Francis Darwin used his father's notes to revise the second edition.

Darwin offered a natural explanation for emotions in humans and other animals and refuted the idea that expressive facial muscles were a special endowment of humans. This was one of the first books to use photographs (heliotypes).

Insectivorous Plants. Charles Darwin. John Murray, London. 1875. 462 pp. 2nd ed. 1888 (337 pp.). The second edition was revised by Francis Darwin.

This book focused on experiments with the insect-eating sundew, *Drosera.* Darwin explained the remarkable adaptations by which the sundew can capture, digest, and use insects for nutrition when it grows in impoverished conditions.

The Effects of Cross- and Self-Fertilisation in the Vegetable Kingdom. Charles Darwin. John Murray, London. 1876. 482 pp. 2nd ed. 1878 (487 pp.).

Darwin demonstrated that the offspring of cross-fertilized plants were more numerous and vigorous than the offspring of self-fertilized ones.

The Different Forms of Flowers on Plants of the Same Species. Charles Darwin. John Murray, London. 1877. 352 pp. 2nd ed. 1878.

Darwin showed that each different flower form within a species is adapted to use the pollen from another form of flower in that species for fertilization. When a flower form is artificially pollinated with its own pollen, less vigorous offspring result.

The Power of Movement in Plants. Charles Darwin (assisted by Francis Darwin). John Murray, London. 1880. 592 pp.

This was an extension of *Climbing Plants.* Darwin discovered that the growing tip of a shoot is light sensitive and that a stem elongates on the side of the shoot away from light. These observations led to the field of plant growth hormones.

The Formation of Vegetable Mould, through the Action of Worms, with Observations on their Habits. Charles Darwin. John Murray, London. 1881. 326 pp.

Darwin calculated that earthworms bring eighteen tons of finely ground soil per acre per year to the surface, thereby aerating and improving the soil for plant growth. This is a classic study in quantitative ecology.

APPENDIX B

Chronology

DATE	EVENT	DARWIN'S AGE
1809 February 12	Born at Shrewsbury, son of Robert Waring Darwin and Susannah née Wedgwood	
1817	Attended Dr. G. Case's day school at Shrewsbury	8
1818	Entered Shrewsbury School	9
1825 October 22	Matriculated at University of Edinburgh	16
1827 October 15	Admitted to Christ's College, Cambridge	18
1831 April 26	Graduated with BA degree	22
1831 August 30	Received invitation to sail on the *Beagle*	22
1831 December 27	HMS *Beagle* sailed from Plymouth	22
1833 January 3	HMS *Beagle* reaches Tierra del Fuego and returns the Fuegians	23
1835 March 26	Heavily bitten by *Triatoma infestans* in the Andes	26
1835 September	Visited Galápagos Islands	26
1836 October 2	Landed at Falmouth, England	27
1837 March 13	Took quarters at 36 Great Marlborough Street, London	28

1837 July	Opened first notebook on transmutation of species	28
1838 October 28	Read Malthus's essay *On Population*	29
1838 November 11	Proposed marriage to Emma Wedgwood, his cousin, and was accepted	29
1839 January 1	Established at 12 Upper Gower Street, London	29
1839 January 24	Elected Fellow of the Royal Society	29
1839 January 29	Married Emma Wedgwood at Maer, Staffordshire	29
1839 August	*Journal of Researches into the Geology and Natural History* [*Voyage of the Beagle*] published	30
1839 December 27	William Erasmus Darwin born	30
1841 March 1	Anne Elizabeth Darwin born	32
1842 May	Wrote sketch of species theory; *Structure and Distribution of Coral Reefs* published	33
1842 September 17	Moved to Down House in Downe, Kent	33
1842 September 23	Mary Eleanor Darwin born (died three weeks later)	33
1843 September 25	Henrietta Emma Darwin born	34
1844 July	Enlarged essay on species theory	35
1844 November	*Geological Observations on Volcanic Islands* published	35
1845 July 9	George Howard Darwin born	36
1845 August	*Journal of Researches* (2nd ed.) published	36
1846 October 1	Began work on barnacles; *Geological Observations on South America* published	37

1847 July 8	Elizabeth Darwin born	38
1848 August 16	Francis Darwin born	39
1850 January 15	Leonard Darwin born	40
1851 April 23	Anne Elizabeth Darwin died	41
1851 May 13	Horace Darwin born	41
1851 June	*Monograph of Fossil Lepadidae* [barnacles] published; *Monograph of* [recent] *Lepadidae* published	42
1853	Awarded Royal Medal of the Royal Society	44
1854 August	*Monograph of Fossil Balanidae* published; *Monograph of* [recent] *Balanidae* published	45
1854 September 9	Began to sort out notes on species	45
1856 May 14	Began to write large work on species	47
1856 December 5	Charles Waring Darwin born	47
1858 June 18	Received from Alfred Russel Wallace a perfect summary of his own theory of evolution by natural selection	49
1858 June 28	Charles Waring Darwin died	49
1858 July 1	Joint paper with Wallace read before Linnean Society of London	49
1858 July 20	Began writing *Origin of Species*	49
1858 August 20	Joint paper with Wallace published	49
1859 November 24	*Origin of Species* published 50 (1250 copies, all sold on first day)	
1860 January 7	*Origin* 2nd ed. published (3000 copies)	51
1861 April	*Origin* 3rd ed. published (2000 copies)	52
1862 May 15	*On the Various Contrivances by which British and Foreign Orchids are Fertilised by Insects* published	53

1864 November 30	Awarded Copley Medal of the Royal Society	55
1866 December 15	*Origin* 4th ed. published (1250 copies)	57
1868 January 30	*Variation of Animals and Plants under Domestication* published (1500 copies)	58
1868 February 20	*Variation of Animals and Plants under Domestication* reprint published (1500 copies)	59
1869 August 7	*Origin* 5th ed. published (2000 copies)	60
1871 February 24	*Descent of Man* published (2500 copies and reprint of 5000)	62
1872 February 19	*Origin* 6th ed. published (3000 copies)	63
1872 November 26	*Expression of Emotions in Man and Animals* published (7000 copies and reprint of 2000)	63
1874 June	*Structure and Distribution of Coral Reefs* 2nd ed. published	65
1874 Autumn	*Descent of Man* 2nd ed. published	65
1875 July 2	*Insectivorous Plants* published	66
1875 September	*Climbing Plants* published	66
1875 end	*Variation of Animals and Plants under Domestication* 2nd ed. published	66
1876 May–June	Began autobiographical sketch	67
1876 December 5	*Effects of Cross- and Self-Fertilisation in the Vegetable Kingdom* published	67
1877 January	*Fertilisation of Orchids* 2nd ed. published	68
1877 July 9	*Different Forms of Flowers on Plants of the Same Species* published	68
1880 November 22	*Power of Movement in Plants* published	71

1881 October 10	*Formation of Vegetable Mould Through the Action of Worms* published	72
1882 April 19	Died at Down House	73
1882 April 26	Buried in Westminster Abby	73

Modified from: Gavin de Beer, ed., *Charles Darwin / Thomas Henry Huxley Autobiographies* (Oxford University Press, London, 1974, 123 pp.), courtesy of Oxford University Press; Gavin de Beer, *Charles Darwin: A Scientific Biography* (Anchor Books, Garden City, NY, 1965, 295 pp.); and Julian Huxley and H. B. D. Kettlewell, *Charles Darwin and His World* (Viking Press, New York, 1965, 144 pp.). First appeared in Tim M. Berra, *Evolution and the Myth of Creationism* (Stanford University Press, Stanford, CA, 1990, 198 pp.).

APPENDIX C

Darwin Online

The Complete Works of Charles Darwin Online. http://darwin-online.org.uk. This website, from the University of Cambridge, has the largest available collection of writings by and about Darwin. It contains lists of his complete publications, thousands of handwritten manuscripts, and a Darwin bibliography and manuscript catalog. There are also over 200 supplementary texts from reference works, reviews, obituaries, biographies, and more.

Darwin Digital Library of Evolution. http://darwinlibrary.amnh.org. The goal of this site, based at the American Museum of Natural History Library, is to make the full range of literature on evolution available online within a historically and topically coherent structure. Charles Darwin's work is central, but this site covers the seventeenth century to the present, and encompasses the history of evolution as a scientific theory with deep roots and broad cultural consequences.

Darwin Correspondence Project. www.darwinproject.ac.uk. This project has located about 14,500 letters to and from Darwin and about 2000 of his correspondents. About 5000 of the entries include transcriptions taken from *The Correspondence of Charles Darwin* (Burkhardt and Smith 1985–).

About Darwin.com. www.aboutdarwin.com/index.html. This very useful, comprehensive website is dedicated to the life and times of Charles Darwin. It was created by David Leff, a self-described amateur scholar of the history of science.

APPENDIX D

Dates

John James Audubon (1785–1851)
Jane Austen (1775–1817)
Benjamin Bynoe (ca. 1804–1865)
Robert Chambers (1802–1883)
Andrew Clark (1826–1893)
Syms Covington (1816–1861)
Anne Elizabeth Darwin (1841–1851)
Caroline Sarah Darwin (1800–1888)
Charles Robert Darwin (1809–1882)
Charles Waring Darwin (1856–1858)
Elizabeth "Bessy" Darwin (1847–1928)
Emily Catherine Darwin (1810–1866)
Emma née Wedgwood Darwin (1808–1896)
Erasmus Alvery Darwin (1804–1881)
Erasmus Darwin (1731–1802)
Francis Darwin (1848–1925)
George Howard Darwin (1845–1912)
Horace Darwin (1851–1928)
Leonard Darwin (1850–1943)
Marianne Darwin (1798–1858)
Mary Eleanor Darwin (1842–1842)
Robert Waring Darwin (1766–1848)
Susan Elizabeth Darwin (1803–1866)
Susannah née Wedgwood Darwin (1764–1817)
William Erasmus Darwin (1839–1914)
Charles Dickens (1812–1870)
Augustus Earle (1793–1838)

Robert FitzRoy (1805–1865)
William Darwin Fox (1805–1880)
King George III (1738–1820)
John Gould (1804–1881)
Robert Edmond Grant (1793–1874)
Asa Gray (1810–1888)
James Gully (1808–1883)
John Stevens Henslow (1796–1861)
Joseph Dalton Hooker (1817–1911)
Alexander von Humboldt (1769–1859)
Thomas Henry Huxley (1825–1895)
Philip Gidley King (1817–1904)
Henrietta Emma née Darwin Litchfield (1843–1929)
John Lubbock (1834–1913)
Charles Lyell (1797–1875)
Thomas Robert Malthus (1766–1834)
Conrad Martens (1801–1878)
Harriet Martineau (1802–1876)
Robert McCormick (1800–1890)
John Murray (1808–1892)
Richard Owen (1804–1892)
George Peacock (1791–1858)
Joseph Priestley (1733–1804)
Gwen Raverat (1885–1957)
Adam Sedgwick (1785–1873)
Herbert Spencer (1820–1903)
John Lort Stokes (1812–1885)
Bartholemew James Sulivan (1810–1890)
Queen Victoria (1819–1901)
Alfred Russel Wallace (1823–1913)
James Watt (1736–1819)
Josiah Wedgwood (1730–1795)
Josiah Wedgwood II (1769–1843)
Josiah Wedgwood III (1795–1880)
John Clements Wickham (1798–1864)
Samuel Wilberforce (1805–1873)

References

THERE ARE HUNDREDS of books and articles about the life of Charles Darwin. The following are a selection from my personal library which I used to prepare this book. Books on my recommended reading list are indicated with an asterisk.

Adams, Alexander B. 1969. *Eternal Quest: The Study of the Great Naturalists*. G. P. Putnam's Sons, New York. 509 pp.

Allan, Mea. 1977. *Darwin and His Flowers: The Key to Natural Selection*. Taplinger, New York. 318 pp.

Anonymous. 1969. *Historical and Descriptive Catalogue of Darwin Memorial at Down House, Downe, Kent*. E. & S. Livingstone, Edinburgh. 30 pp.

Armstrong, Patrick. 1985. *Charles Darwin in Western Australia*. University of Western Australia Press, Nedlands, Australia. 75 pp.

Atkins, Hedley. 1974. *Down: The Home of the Darwins*. Royal College of Surgeons of England, London. 127 pp.

Barloon, Thomas J. and Russell Noyes, Jr. 1997. Charles Darwin and panic disorder. *Journal of the American Medical Association* 277(2): 138–41.

Barlow, Nora. (Ed.). 1946. *Charles Darwin and the Voyage of the* Beagle. Philosophical Library, New York. 279 pp.

———. (Ed.). 1967. *Darwin and Henslow: The Growth of an Idea; Letters 1831–1860*. Bentham-Moxon Trust/John Murray, London. 251 pp.

*———. (Ed.). 1958. *The Autobiography of Charles Darwin 1809–1882: With Original Omissions Restored*. W. W. Norton, New York. 253 pp.

Barrett, Paul H. (Ed.). 1977. *The Collected Papers of Charles Darwin*. 2 vols. University of Chicago Press, Chicago.
Barrett, Paul, Peter J. Gautrey, Sandra Herbert, David Kohn, and Sydney Smith. (Eds.). 1987. *Charles Darwin's Notebooks, 1836–1844*. Cornell University Press, Ithaca, NY. 747 pp.
Barrett, Paul H., Donald J. Weinshank, and Timothy T. Gottleber. (Eds.). 1981. *A Concordance to Darwin's* Origin of Species, *First Edition*. Cornell University Press. Ithaca, NY. 834 pp.
Berra, Tim M. 1980. Charles Darwin: What else did he write? *American Biology Teacher* 42(8): 489–92.
———. 1990. *Evolution and the Myth of Creationism*. Stanford University Press, Stanford, CA. 198 pp.
———. 2006. Art, ichthyology, Charles Darwin and the Northern Territory of Australia. *The Beagle, Records of the Museums and Art Galleries of the Northern Territory* 22:91.
———. 2008. Charles Darwin's paradigm shift. *The Beagle, Records of the Museums and Art Galleries of the Northern Territory* 24 (forthcoming).
Berry, R. J. 1982. Charles Darwin: A Commemoration 1882–1982. *Biological Journal of the Linnean Society* 17(1): 1–135.
Bowlby, John. 1990. *Charles Darwin: A New Life*. W. W. Norton, New York. 511 pp.
Bowler, Peter J. 1984. *Evolution: The History of an Idea*. University of California Press, Berkeley. 412 pp.
———. 1990. *Charles Darwin: The Man and His Influence*. Cambridge University Press, Cambridge. 250 pp.
———. 2002. Climb Chimborazo and see the world. *Science* 298: 63–64.
Brackman, Arnold C. 1980. *A Delicate Arrangement: The Strange Case of Charles Darwin and Alfred Russel Wallace*. Times Books, New York. 369 pp.
Bradford, Gamaliel. 1926. *Darwin*. Houghton Mifflin, Boston. 315 pp.
Brent, Peter. 1981. *Charles Darwin: A Man of Enlarged Curiosity*. W. W. Norton, New York. 536 pp.
*Browne, Janet. 1995. *Charles Darwin: A Biography*. Vol. 1, *Voyaging*. Princeton University Press, Princeton, NJ. 605 pp.

*———. 2002. *Charles Darwin: A Biography*. Vol. 2, *The Power of Place*. Knopf, New York. 591 pp.

———. 2006. *Darwin's Origin of Species: A Biography*. Atlantic Monthly Press, New York. 174 pp.

Bunting, James. 1974. *Charles Darwin: A Biography*. Bailey Brothers & Swinfen, Folkestone, UK. 126 pp.

Burkhardt, Frederick and Sydney Smith. (Eds.). 1985–. *The Correspondence of Charles Darwin*. 15 vols. Cambridge University Press, Cambridge. [This project is expected to include 32 volumes and be completed about 2025.]

Camerini, Jane R. (Ed.). 2002. *The Alfred Russel Wallace Reader*. Johns Hopkins University Press, Baltimore. 219 pp.

Chapman, Roger G. and Cleveland T. Duval. (Eds.). 1982. *Charles Darwin 1809–1882: A Centennial Commemorative*. Nova Pacifica, Wellington, New Zealand. 376 pp.

Clark, Ronald W. 1984. *The Survival of Charles Darwin: A Biography of a Man and an Idea*. Random House, New York. 449 pp.

Clodd, Edward. 1903. *Pioneers of Evolution from Thales to Huxley*. Grant Richards, London. 250 pp.

*Colp, Ralph Jr. 1977. *To Be an Invalid: The Illness of Charles Darwin*. University of Chicago Press, Chicago. 285 pp.

Darwin, Charles. 1909. *The Foundations of the Origin of Species: Two Essays Written in 1842 and 1844*. Cambridge University Press, Cambridge. 263 pp. (Reprint 1969, Kraus Reprint, New York.)

Darwin, Francis. (Ed.). 1897. *The Life and Letters of Charles Darwin: Including an Autobiographical Chapter*. 2 vols. D. Appleton, New York.

*De Beer, Gavin. 1965. *Charles Darwin: A Scientific Biography*. Natural History Library. Anchor Books, Garden City, NY. 295.

———. (Ed.). 1974. *Autobiographies: Charles Darwin / Thomas Henry Huxley*. Oxford University Press, London. 123 pp.

Desmond, Adrian. 1997. *Huxley: From Devil's Disciple to Evolution's High Priest*. Perseus Books, Reading, MA. 820 pp.

*Desmond, Adrian and James Moore. 1991. *Darwin: The Life of a Tormented Evolutionist*. Warner Books, New York. 808 pp.

Desmond, Adrian, James Moore, and Janet Browne. 2007. *Charles Darwin*. Oxford University Press, Oxford. 139 pp.
Dibner, Bern. 1960. *Darwin of the* Beagle. Burndy Library, Norwalk, CT. 73 pp.
Dobson, Jessie. 1971. *Charles Darwin and Down House*. Churchill Livingstone, Edinburgh. 16 pp.
Dorsey, George A. 1927. *The Evolution of Charles Darwin*. Doubleday, Page, Garden City, NY. 300 pp.
Eiseley, Loren. 1961. *Darwin's Century: Evolution and the Men Who Discovered It*. Anchor Books, Garden City, NY. 378 pp.
———. 1979. *Darwin and the Mysterious Mr. X*. E. P. Dutton, New York. 278 pp.
Eldredge, Niles. 2005. *Darwin: Discovering the Tree of Life*. W. W. Norton, New York. 256 pp.
*Engel, Leonard. (Ed.). 1962. *The Voyage of the* Beagle. Natural History Library. Anchor Books, Garden City, NY. 524 pp.
Ferguson, B. J. 1971. *Syms Covington of Pambula*. Imlay District Historical Society, Merimbula, Australia. 17 pp.
Freeman, R. B. 1977. *The Works of Charles Darwin: An Annotated Bibliographical Handlist*. 2nd ed., revised and enlarged. Archon Books, Hamden, CT. 235 pp.
———. 1978. *Charles Darwin: A Companion*. Archon Books, Hamden, CT. 309 pp.
Gelb, Michael H. and Wim G. J. Hol. 2002. Drugs to combat tropical protozoan parasites. *Science* 297: 343–44.
George, Wilma. 1964. *Biologist Philosopher: A Study of the Life and Writing of Alfred Russel Wallace*. Abelard-Schuman, London. 320 pp.
Ghiselin, Michael T. 1969. *The Triumph of the Darwinian Method*. University of California Press, Berkeley. 287 pp.
Gillespie, Neal C. *Charles Darwin and the Problem of Creation*. University of Chicago Press, Chicago. 201 pp.
Gould, Stephen Jay. 1986. Knight takes bishop? The facts about the great Wilberforce-Huxley debate don't always fit the legend. *Natural History* 95(5): 18–33.
Gregor, Arthur S. 1967. *Charles Darwin*. Angus & Robertson, London. 189 pp.

Gribbin, John and Mary Gribbin. 2003. *FitzRoy: The Remarkable Story of Darwin's Captain and the Invention of the Weather Forecast*. Yale University Press, New Haven, CT. 336 pp.

Gruber, Howard. 1974. *Darwin on Man: A Psychological Study of Scientific Creativity Together with Darwin's Early Unpublished Notebooks*. Transcribed and annotated by Paul H. Barrett. E. P. Dutton, New York. 495 pp.

Healey, Edna. 1986. *Wives of Fame: Mary Livingstone, Jenny Marx, Emma Darwin*. Sidgwick & Jackson, London. 210 pp.

Herbert, Sandra. 2005. *Charles Darwin, Geologist*. Cornell University Press, Ithaca, NY. 485 pp.

Hosler, Jay. 2003. *Sandwalk Adventures*. Active Synapse, Columbus, OH. 159 pp.

Howard, Jonathan. 1982. *Darwin*. Oxford University Press, Oxford. 101 pp.

Hull, David L. 1973. *Darwin and His Critics: The Reception of Darwin's Theory of Evolution by the Scientific Community*. University of Chicago Press, Chicago. 473 pp.

Huxley, Julian and H. B. D. Kettlewell. 1965. *Charles Darwin and His World*. Viking Press, New York. 144 pp.

Huxley, Leonard. 1900. *Life and Letters of Thomas Henry Huxley*. 2 vols. Macmillan, London.

Irvine, William. 1956. *Apes, Angels & Victorians: A Joint Biography of Darwin & Huxley*. Readers Union / Weidenfeld & Nicolson, London. 278 pp.

Jones, Steve. 2000. *Darwin's Ghost*. Ballantine Books, New York. 377 pp.

Karp, Walter. 1968. *Charles Darwin and the* Origin of Species. American Heritage Publishing, New York. 153 pp.

*Keynes, Randal. 2001. *Annie's Box: Charles Darwin, His Daughter and Human Evolution*. Fourth Estate, London. 331 pp. (US title: *Darwin, His Daughter & Human Evolution*.)

Keynes, Richard Darwin. (Ed.). 1979. *The* Beagle *Record: Selections from the Original Pictorial Records and Written Accounts of the Voyage of H. M. S.* Beagle. Cambridge University Press, Cambridge. 409 pp.

———. 2003. *Fossils, Finches, and Fuegians.* Oxford University Press, Oxford. 428 pp.

Kohn, David. (Ed.). 1985. *The Darwin Heritage.* Princeton University Press, Princeton, NJ. 1,138 pp.

McDonald, Roger. 1998. *Mr. Darwin's Shooter.* Atlantic Monthly Press, New York. 365 pp.

Manier, Edward. 1978. *The Young Darwin and His Cultural Circle.* D. Reidel, Boston. 242 pp.

Marks, Richard Lee. 1991. *Three Men of the* Beagle. Alfred A. Knopf, New York. 256 pp.

Marshall, A. J. 1970. *Darwin and Huxley in Australia.* Hodder & Staughton, Sydney. 142 pp.

Mayr, Ernst. 2000. Darwin's influence on modern thought. *Scientific American* 283(1): 78–83.

Meacham, Standish. 1970. *Lord Bishop: The Life of Samuel Wilberforce 1805–1873.* Harvard University Press, Cambridge, MA. 328 pp.

Meadows, Jack. 1989. *The Great Scientists.* Oxford University Press, Oxford. 256 pp.

Mellersh, H. E. L. 1968. *Fitzroy of the* Beagle. Mason & Lipscomb, New York. 308 pp.

Miller, Jonathan and Borin Van Loon. 1982. *Darwin for Beginners.* Pantheon Books, New York. 176 pp.

Milner, Richard. 1994. *Charles Darwin: Evolution of a Naturalist.* Universities Press, Hyderabad, India. 158 pp.

———. 1995. Charles Darwin: The last portrait. *Scientific American* 273(5): 78–79.

———. (Ed.). 2005. Darwin & evolution. Special issue, *Natural History* 114(9). [Multiple authors with articles on the historical Darwin and Darwinism today.]

Moore, James, R. 1989. Of love and death: Why Darwin "gave up Christianity." In *History, Humanity and Evolution: Essays for John C. Greene,* James R. Moore (Ed.), 195–229. Cambridge University Press, Cambridge.

———. 1994. *The Darwin Legend.* Baker Books, Grand Rapids, MI. 218 pp.

*Moorehead, Alan. 1969. *Darwin and the* Beagle. Hamish Hamilton, London. 280 pp.

Nicholas, F. W. and J. M. Nicholas. 1989. *Charles Darwin in Australia*. Cambridge University Press, Cambridge. 175 pp.
Nichols, Peter. 2003. *Evolution's Captain*. Harper Collins, New York. 336 pp.
Pauly, Daniel. 2004. *Darwin's Fishes: An Encyclopedia of Ichthyology, Ecology and Evolution*. Cambridge University Press, Cambridge. 340 pp.
Peattie, Donald Culross. 1936. *Green Laurels: The Lives and Achievements of the Great Naturalists*. Literary Guild, New York. 368 pp.
Pekin, L. B. 1938. *Darwin*. Stackpole Sons, New York. 110 pp.
Quammen, David. 2006. *The Reluctant Mr. Darwin*. W. W. Norton, New York. 304 pp.
Raby, Peter. 2001. *Alfred Russel Wallace: A Life*. Princeton University Press. Princeton, NJ. 340 pp.
Ralling, Christopher. (Ed.). 1978. *The Voyage of Charles Darwin*. British Broadcasting Corporation, London. 183 pp.
*Raverat, Gwen. 1971. *Period Piece: A Cambridge Childhood*. Faber & Faber, London. 282 pp.
Ruse, Michael. 1979. *The Darwinian Revolution: Science Red in Tooth and Claw*. University of Chicago Press, Chicago. 320 pp.
Sears, Paul B. 1950. *Charles Darwin: The Naturalist as a Cultural Force*. Charles Scribner's Sons. New York, NY. 124 pp.
Sloan, Pat. 1960. The myth of Darwin's conversion. *Humanist* 75(3): 70–72.
———. 1965. Demythologizing Darwin. *Humanist* 80(4): 106–10.
Stone, Irving. 1980. *The Origin: A Biographical Novel of Charles Darwin*. Doubleday, Garden City, NY. 743 pp. [Reviewed in 1981 by Stephen Jay Gould, Darwin novelized, *Science* 211(4479): 270–71.]
Stott, Rebecca. 2003. *Darwin and the Barnacle*. W.W. Norton, New York. 309 pp.
Subercaseaux, Benjamin. 1954. *Jemmy Button*. Macmillan, New York. 382 pp.
Sulloway, Frank J. 1982. Darwin and his finches: The evolution of a legend. *Journal of the History of Biology* 15(1): 1–53.
———. 1982. Darwin's conversion: The *Beagle* voyage and its aftermath. *Journal of the History of Biology* 15(3): 325–96.

———. 1984. Darwin and the Galápagos. *Biological Journal of the Linnean Society* 21: 29–59.

Thompson, Keith S. 1995. *HMS* Beagle*: The Story of Darwin's Ship*. W. W. Norton, New York. 320 pp.

Tobias, Phillip V. 1972. Darwin, "Descent" and Disease. *Transactions of the Royal Society of South Africa* 40(4): 239–60.

Turrill, W. B. 1963. *Joseph Dalton Hooker: Botanist, Explorer, and Administrator*. Thomas Nelson & Sons, London. 228 pp.

Wallace, Alfred Russel. 1905. *My Life: A Record of Events and Opinions*. 2 vols. Dodd, Mead, New York.

Ward, Henshaw. 1927. *Charles Darwin and the Theory of Evolution*. New Home Library, New York. 472 pp.

Weiner, Johnathan. 1994. *The Beak of the Finch*. Alfred A. Knopf, New York. 332 pp.

West, Geoffrey. 1937. *Charles Darwin: The Fragmentary Man*. George Routledge & Sons, London. 351 pp.

White, Michael and John Gribbin. 1997. *Darwin: A Life in Science*. Plume Books, New York. 322 pp.

*Wilson, Louise. (Ed.). 2000. *Down House: The Home of Charles Darwin*. English Heritage, London. 48 pp. [Text by Solene Morris and Louise Wilson.]

Wrangham, R. 1979. The Bishop of Oxford: Not so soapy. *New Scientist* 83(1167): 450–51.

Zimmer, Carl. 2001. *Evolution: The Triumph of an Idea*. HarperCollins, New York. 364 pp.

Index

Page numbers preceded by "G" refer to the illustration gallery.

Angraecum sesquipedale, 72
apes, 74
Audubon, John James, 8
Australia, 18, 21, 31, 86

barnacles, 52–53, 56–57, 89, G15
Beagle, H.M.S., 13–16, 20–25, 29, 31, 33, G7; mission, 14
Bynoe, Benjamin, 21

Cambridge University, 9–10, 80
Chagas's disease, 26
Chambers, Robert, 50–51
Clark, Andrew, 82
Copley Medal, 80
coral reefs, 31, 33
Coral Reefs, 41, 88–89
Covington, Syms, 18

Darwin, Anne Elizabeth, 40–41, 54–56, 81
Darwin, Caroline Sarah, 5, 39
Darwin, Charles Robert: *Beagle* invitation, 13; birth, 4; Cambridge University, 9, 83; correspondence, 45–46, 99; crossing the equator, 20; death, 82–83; family tree, 61; finances, 39, 58; funeral, 82–83; Galápagos Islands, 29–31; graduation, 12; illness, 26–27, 37, 40, 49, 53, 82; marriage, 39; medical school, 7–9; *Megatherium* discovery, 19, 34; notebooks, 36, 40, G9; portrait at age seven, G4; portrait, wedding, G11; religious views, 9, 40, 55, 81; return to England, 34; seasickness, 15; sketch of species theory, 41, 50–51; tree of life, 36
Darwin, Charles Waring, 60, 62
Darwin, Elizabeth ("Bessy"), 53, 57, 83
Darwin, Emily Catherine, 5, G4
Darwin, Emma (née Wedgwood), 9, 39–40, 47, 49, 51, 55, 57–58, 62, 77, 80, G11
Darwin, Erasmus, 2, 4, 9, G1
Darwin, Erasmus Alvery, 5–7, 9, 35, 37
Darwin, Francis, 53, 57, 78–79, 81, 83, 85, 91

Darwin, George Howard, 50
Darwin, Henrietta Emma, 50, 54, 57, 60, 83, 85
Darwin, Horace, 56–57, 79
Darwin, Leonard, 54–55, 57
Darwin, Marianne, 5
Darwin, Mary Eleanor, 42
Darwin, Northern Territory, Australia, x, 31, 86
Darwin, Robert Waring, 3, 6–7, 9, 13, 39, 53, 58
Darwin, Susan Elizabeth, 5
Darwin, Susannah (née Wedgwood), 4, 6, G2
Darwin, William Erasmus, 40–41
"Darwin's Bulldog," 46, 58
Darwin's finches, 29, 31–32
Descent of Man and Selection in Relation to Sex, 73, 90–91
Dickens, Charles, 34
Different Forms of Flowers on Plants of the Same Species, 78, 91–92
Downe, village of, 42
Down House, 42–51, 57, 62, 73, 76–79, 82, 85–86, G5, G12–G14

Earle, Augustus, 20–21, G6
earthquake, 26, 28
earthworms, 79
Edinburgh University, 7–9
Edmonston, John, 8
Effects of Cross- and Self-Fertilization in the Vegetable Kingdom, 77, 91
evolution, 19, 37, 68–69, 74, 90
Expression of Emotions in Man and Animals, 75–76, 91
extinction, 19

FitzRoy, Robert, 13–14, 19–20, 25, 33, 44, 70, G5
Formation of Vegetable Mould Through the Action of Worms With Observations on Their Habits, 78–80, 92
Fox, William Darwin, 9–10
Fuegians, 21–24

Galápagos: finches, 31–32; iguanas, 31, G8; mockingbirds, 29, 31; tortoise, 29–30
Galápagos Islands, 29–31, 86, G8
Geochelone elephantopus, 30
Geological Observations on South America, 52, 88–89
Geology of the Voyage of the Beagle, 37, 41, 88–89
Geospiza magnirostris, 32
Gould, John, 31–32
Grant, Robert Edmond, 7
Gray, Asa, 47, 59, 62, 64, 72, 78
Gully, James, 53–54

Henslow, John Stevens, 11–12, 18, 33–34
Hooker, Joseph Dalton, 17, 45–46, 50, 52, 59, 62–64, 83
Humboldt, Alexander von, 12
Huxley, Thomas Henry, 46, 56, 58–59, 69–71, 81, 83

iguana, marine and land, 31, G8
Insectivorous Plants, 76, 91

Journal of Researches (*Voyage of the Beagle*), 35, 66, 87–88

King, Philip Gidley, 15, 31

Lady Hope, 85
Linnean Society, 63–64, 72
Litchfield, Henrietta (née Darwin), 50, 85
Lubbock, John, 59, 83
Lyell, Charles, 16–17, 27, 34, 46, 59–60, 62, 64, 83

Maer Hall, Staffordshire, 6, 13
Malthus, Thomas Robert, 40, 59
marriage, pros and cons of, 37–38
Martens, Conrad, 21–22, 24
Martineau, Harriet, 37
Matthews, Richard, 22, 24
McCormick, Robert, 20
Megatherium, 19, 34
midshipmen's berth, painting of (Earle), G6
Mount, Shrewbury, 6–7, 13
Movements and Habits of Climbing Plants, 72, 89, 92
Murray, John, 66, 73
mutation, 69

natural selection, 37, 41, 52, 59–60, 63, 68–69, 86
Nesomimus sp., 29

On the Various Contrivances by which Orchids are Fertilized by Insects, 72, 90
orchids, 72, 77
Origin of Species, 65–67, 71–74, 86, 89–91
Owen, Fanny, 9, 11, 18
Owen, Richard, 34, 67, 70

pigeons, 57, 73, G16
Power of Movement in Plants, 78, 90, 92
Principles of Geology, 16–17

Raverat, Gwen, 50
Royal Society, 39, 80, 83

Sedgwick, Adam, 12
slavery: Darwin on, 19; depicted, 4, 21
Spencer, Herbert, 73, 90
Stokes, John Lort, 15, 31
Sulivan, Bartholemew James, 21
"survival of the fittest," 73, 90

Tierra del Fuego, 21–25
transmutation, 19, 36–37, 40, 74
tree of life, 36–37
Triatoma infestans, 26, 29
Trypanosoma cruzi, 26, 29

uniformitarianism, 16–17

Variation of Animals and Plants under Domestication, 73, 90
Vestiges of the Natural History of Creation, 50
Victoria, Queen, 80, 82
volcanic eruption, 26, 28
Voyage of the Beagle, 25, 32, 35, 37, 66, 87–88; map, 33

Wallace, Alfred Russel, 59–60, 62–64, 66, 83
"water cure," 53–54
Wedgwood, Emma, 8–9, 39
Wedgwood, Josiah, 2, 4, G2
Wedgwood II, Josiah, 6, 8, 13, 39, 55, G2
Wedgwood III, Josiah, 39
Wedgwood, Susannah, 4, 6, G2
Wedgwood ware, G3
Wickham, John Clements, 20–21, 28
Wilberforce, Samuel, 70–71

Xanthopan morganii praedicta, 72

Zoology of the Voyage of the Beagle, 37, 87
Zoonomia, 2

About the Author

TIM M. BERRA is professor emeritus of evolution, ecology, and organismal biology at the Ohio State University, where he has been since 1972. He is also a research associate at the Museum and Art Gallery of the Northern Territory in Darwin, NT, Australia. Berra is a two-time Fulbright Fellow to Australia and has spent about eight of the past thirty-nine years doing fieldwork on fishes throughout Australia. He has taught at the University of Papua New Guinea, and has been a visiting professor at the University of Concepción, Chile, and the University of Otago, New Zealand. He received a PhD in biology from Tulane University in 1969. He and his wife reside in a home with 15,000 books in its library, on a twenty-acre property in Amish country near Bellville, Ohio, surrounded by a lake and stream with forty-four species of fishes.

Also by Tim M. Berra

William Beebe: An Annotated Bibliography

An Atlas of Distribution of the Freshwater Fish Families of the World

Evolution and the Myth of Creationism

A Natural History of Australia

Freshwater Fish Distribution